About IFPRI

The International Food Policy Research Institute (IFPRI), established in 1975, provides research-based policy solutions to sustainably reduce poverty and end hunger and malnutrition. The Institute conducts research, communicates results, optimizes partnerships, and builds capacity to ensure sustainable food production, promote healthy food systems, improve markets and trade, transform agriculture, build resilience, and strengthen institutions and governance. Gender is considered in all of the Institute's work. IFPRI collaborates with partners around the world, including development implementers, public institutions, the private sector, and farmers' organizations. IFPRI is a member of the CGIAR Consortium.

About IFPRI's Peer Review Process

IFPRI books are policy-relevant publications based on original and innovative research conducted at IFPRI. All manuscripts submitted for publication as IFPRI books undergo an extensive review procedure that is managed by IFPRI's Publications Review Committee (PRC). Upon submission to the PRC, the manuscript is reviewed by a PRC member. Once the manuscript is considered ready for external review, the PRC submits it to at least two external reviewers who are chosen for their familiarity with the subject matter and the country setting. Upon receipt of these blind external peer reviews, the PRC provides the author with an editorial decision and, when necessary, instructions for revision based on the external reviews. The PRC reassesses the revised manuscript and makes a recommendation regarding publication to the director general of IFPRI. With the director general's approval, the manuscript enters the editorial and production phase to become an IFPRI book.

Food Security in a World of Natural Resource Scarcity

The Role of Agricultural Technologies

Mark W. Rosegrant, Jawoo Koo, Nicola Cenacchi, Claudia Ringler,
Richard Robertson, Myles Fisher, Cindy Cox, Karen Garrett,
Nicostrato D. Perez, and Pascale Sabbagh

A Peer-Reviewed Publication

International Food Policy Research Institute
Washington, DC

International Food Policy Research Institute
2033 K Street, NW
Washington, DC 20006-1002, USA
Telephone: +1-202-862-5600
www.ifpri.org

DOI: http://dx.doi.org/10.2499/9780896292079

Library of Congress Cataloging-in-Publication Data

Rosegrant, Mark W.
  Food security in a world of natural resource scarcity : the role of agricultural technologies / Mark W. Rosegrant, Jawoo Koo, Nicola Cenacchi, Claudia Ringler, Richard Robertson, Myles Fisher, Cindy Cox, Karen Garrett, Nicostrato D. Perez, Pascale Sabbagh. —Edition 1.
     pages   cm
  Includes bibliographical references.
  ISBN 978-0-89629-847-7 (alk. paper)
  1. Alternative agriculture.   2. Food security.   3. Natural resources—Management.   4. Crop yields.   5. Agriculture—Mathematical models.
  I. International Food Policy Research Institute.   II. Title.
  S494.5.A65R67   2014
  333.79′66—dc23   2013050175

Cover design: Deirdre Launt
Project manager: Patricia Fowlkes
Book layout: Princeton Editorial Associates Inc., Scottsdale, Arizona

Contents

The appendixes for this book are available online at http://www.ifpri.org/publication/
food-security-world-natural-resource-scarcity.

Tables, Figures, and Boxes

Tables

Boxes

Abbreviations and Acronyms

A1B   greenhouse gas emissions scenario that assumes fast economic growth, a population that peaks mid-century, and the development of new and efficient technologies, along with a balanced use of energy sources

ASI   anthesis-to-silking interval

CIMMYT Centro Internacional de Mejoramiento de Maíz y Trigo (International Maize and Wheat Improvement Center)

CSIRO   Commonwealth Scientific and Industrial Research Organisation's general circulation model

DSSAT   Decision Support System for Agrotechnology Transfer

EU   European Union

FPU   food-producing unit

GHI   Global Hunger Index

GPS   global positioning system

IFPRI   International Food Policy Research Institute

IMPACT International Model for Policy Analysis of Agricultural Commodities and Trade

IPCC   Intergovernmental Panel on Climate Change

IRRI   International Rice Research Institute

ISFM   integrated soil fertility management

MIROC   Model for Interdisciplinary Research on Climate

NUE	nitrogen-use efficiency
OA	organic agriculture
OCR	organic-to-conventional crop yield ratio
PA	precision agriculture
PAW	pathogen, arthropod, weed
R&D	research and development
RCP	Representative Concentration Pathway
SRES	Special Report on Emissions Scenarios
SSA	Africa south of the Sahara
SSP	Shared Socioeconomic Pathway

Foreword

Addressing the challenges of climate change, rising long-term food prices, and poor progress in improving food security will require increased food production without further damage to the environment. Accelerated investments in agricultural research and development will be crucial to supporting food production growth. The specific set of agricultural technologies that should be brought to bear remains unknown, however. At the same time, the future technology mix will have major impacts on agricultural production, food consumption, food security, trade, and environmental quality in developing countries.

Technology options are many, but transparent evidence-based information to support decisions on the potential of alternative technologies is relatively scarce. This is no longer a question of low- versus high-income countries but one of the planet: how do we achieve food security in a world of growing scarcity? Thus, a key challenge for our common future will be how we can grow food sustainably—meeting the demands of a growing population without degrading our natural resource base.

This is the question that this book sets out to address, combining spatially disaggregated crop models linked to economic models to explore the impacts on agricultural productivity and global food markets of 11 alternative agricultural technologies as well as selected technology combinations for maize, rice, and wheat, the world's key staple crops. The book uses a groundbreaking modeling approach that combines comprehensive process-based modeling of agricultural technologies globally with sophisticated global food demand, supply, and trade modeling.

Across the three crops, the largest yield gains, in percentage terms, are in Africa, South Asia, and parts of Latin America and the Caribbean. The book finds wide heterogeneity in yield response, making it important to target specific technologies to specific regions and countries. Heat-tolerant varieties, no-till, nitrogen-use efficiency, and precision agriculture are technologies with particularly great potential for yield improvement in large parts of the world. Moving these technologies forward will require institutional, policy, and investment advances in many areas. Although getting there will not be easy or quick, we must move ahead. The cost of not taking any action could be dramatic for the world's food-insecure.

Shenggen Fan
Director General, IFPRI

Acknowledgments

We thank CropLife International, the U.S. State Department, and the CGIAR Research Program on Policies, Institutions, and Markets for funding this work. We appreciate the guidance and insights from the Study Advisory Panel members for the project that led to this book, in particular, Timothy Benton, Jason Clay, Elisio Contini, Swapan Datta, Lindiwe Sibanda, and Ren Wang. We are grateful for the research support and assistance of Mandy Ewing and Divina Gracia Pagkaliwagan Rodriguez. We also thank Daniel Mason-D'Croz and Prapti Bhandary for their help with the IMPACT model. Xiuqin Bai and Xin Sun of the Department of Plant Pathology, Kansas State University, and Robert Hijmans of the Department of Environmental Science and Policy, University of California, Davis, contributed to the pest prevalence maps. We also acknowledge the administrative and formatting support of Lorena Danessi.

Chapter 1

Introduction

The International Food Policy Research Institute (IFPRI) business-as-usual projections of agricultural supply and demand anticipate a rise in food prices of most cereals and meats, reversing long-established downward trends. Between 2005 and 2050, food prices for maize, rice, and wheat are projected to increase by 104, 79, and 88 percent, respectively, while those for beef, pork, and poultry will rise by 32, 70, and 77 percent, respectively. Moreover, the number of people at risk of hunger in the developing world will grow from 881 million in 2005 to more than a billion people by 2050 (IFPRI International Model for Policy Analysis of Agricultural Commodities and Trade [IMPACT] baseline, Model for Interdisciplinary Research on Climate [MIROC] A1B scenario[1] used in this book). More recent modeling efforts that use nine agricultural models, including both general equilibrium and partial equilibrium models, project that food price increases out to 2050 will be more moderate under climate change, with the IMPACT results in the medium range of price increases. Our results indicate increases in the real price of maize of 40–45 percent in 2050 and in the price of wheat and rice of 20–25 percent under climate change relative to a no–climate change scenario, using the Intergovernmental Panel on Climate Change (IPCC) Fifth Assessment with Representative Concentration Pathway (RCP) 8.5 and Shared Socioeconomic Pathway (SSP) 2 scenario[2] (Nelson et al. 2013).

Both demand and supply factors will drive price increases. Population and regional economic growth will fuel increased growth in demand for food. Rapid growth in demand for meat and milk will put pressure on prices for maize, coarse grains, and meats. World food markets will tighten, adversely affecting poor consumers. The substantial increase in food prices will cause relatively slow growth in calorie consumption, with both direct price impacts on

---

1 A1B is the greenhouse gas emissions scenario that assumes fast economic growth, a population that peaks mid-century, and the development of new and efficient technologies, along with a balanced use of energy sources.

2 SSP2 approximates medium growth rates for population and gross domestic product, and RCP8.5 projects a high temperature increase of 4.5°C by 2100.

the food insecure and indirect impacts through reductions in real incomes for poor consumers who spend a large share of their income on food. This in turn contributes to slow improvement in food security, particularly in South Asia and Africa south of the Sahara (SSA). As productivity growth is insufficient to meet effective demand in much of the developing world, net food imports are expected to increase significantly for the group of developing countries (Rosegrant, Paisner, and Meijer 2003).

In the longer term, adverse impacts from climate change are expected to raise food prices further and dampen developing-country food demand translating into direct increases in malnutrition levels, with often irreversible consequences for young children (Nelson et al. 2010). Climate change could decrease maize yields by 9–18 percent depending on climate change scenario, cropping system (rainfed or irrigated), and whether the carbon fertilization effect is included; rice yields could drop by 7–27 percent; and wheat yields would be particularly affected, sharply declining by 18–36 percent by 2050, compared to a scenario with no climate change (Nelson et al. 2009).

Furthermore, there is now a growing understanding that natural resources are beginning, to a substantial degree, to limit economic growth and human well-being goals (Ringler, Bhaduri, and Lawford 2013). The effects of natural resource scarcity have been described in many recent scientific publications, such as the reports of the IPCC (IPCC, various years), the Millennium Ecosystem Assessment (MA and WRI 2005), and the "Planetary Boundaries" paper (Rockström et al. 2009), and are being debated in many intergovernmental venues that focus on the development of the Sustainable Development Goals that would replace the Millennium Development Goals in 2015 (SDSN 2013).

Rapidly rising resource scarcity of water and increasingly of land will add further constraints on food production growth. At the same time, bioenergy demand will continue to compete with food production for land and water resources despite recent reviews of biofuel policies in the European Union (EU) and the United States (Rosegrant, Fernandez, and Sinha 2009; Rosegrant, Tokgoz, and Bhandary 2013). Given the continued growth of competing demands on water and land resources from agriculture, urbanization, industry, and power generation, food production increases through large expansion into new lands will be unlikely. Land expansion would also entail major environmental costs and damage remaining forest areas and related ecosystem services (Rosegrant et al. 2001; Alston, Beddow, and Pardey 2009; Rosegrant, Fernandez, and Sinha 2009; Foley et al. 2011; Pretty, Toulmin, and Williams 2011; Balmford, Green, and Phalan 2012). Therefore, greater food production will largely need to come from higher productivity rather than from a net increase in cropland area.

Accelerated investments in agricultural research and development (R&D) will be crucial to slow or reverse these recent trends. For the most part, growth rates of yields for major cereals have been slowing in direct response to the slowdown of public agricultural R&D spending during the 1990s (Alston, Beddow, and Pardey 2009; Ainsworth and Ort 2010). However, developing-country spending has picked up over the past decade, mostly driven by China and India (Beintema et al. 2012). It is uncertain whether R&D spending will continue to grow, but more is needed to sustain the growth of agricultural productivity.

Accelerated investments to support improved agricultural technologies and practices will be crucial to slow and reverse these trends, increase productivity, and meet the growing food demands in an environmentally sustainable way. The future choices and adoption of agricultural technologies will fundamentally influence not only agricultural production and consumption but also trade and environmental quality in developing countries. These choices will have implications for water, land, and energy resources, as well as for climate change adaptation and mitigation. The effectiveness of different agricultural technologies is often a polarized debate. At one end of the spectrum, advocates of intensive agriculture assume that massive investments in upstream agricultural science (including biotechnology and genetic modification) are needed for rapid growth of agricultural production, together with high levels of agricultural inputs, such as fertilizer, pesticides, and water. At the other end of the spectrum, advocates of low-input agriculture emphasize the role of organic and low-input agriculture and crop management improvement through water harvesting, no-till, and soil fertility management in boosting future yield growth. In the middle of all this are almost one billion food-insecure people whose food and nutritional security will depend on agricultural technology strategy decisions undertaken by governments and private investors.

Goals of This Study

Given the many options and lack of direction, significant improvements in the quality, transparency, and objectivity of strategic investment decisions about agricultural technologies and associated policies are urgently needed. This book seeks to fill this gap. It contributes to the understanding of future benefits from alternative agricultural technologies by assessing future scenarios for the potential impact and benefits of these technologies on yield growth and production, food security, the demand for food, and agricultural trade. The future pathways for agricultural technology generation, adoption, and use will have major effects on agricultural production, food consumption, food security, trade, and environmental quality in developing countries. Comprehensive impact scenario analysis can contribute to understanding the role of alternative technologies considered in the context of broader agricultural sector policies and investment strategies.

The overall objective of this book is to identify the future impact of alternative agricultural technology strategies for food supply, demand, prices, and food security for the three key staple crops: maize, rice, and wheat. We have done this by (1) analyzing the potential payoffs (yield growth and food security) of alternative agricultural technologies at global and regional levels, taking into account the spatial variability of crop production, climate, soil, and projected climate change; and (2) assessing the market-level consequences of broad adoption of yield-enhancing crop technologies at regional and global scales, as mediated through impacts on commodity markets and trade. We focus our analysis of agricultural technologies on countries and regions that are at risk of hunger (as measured by the 2013 Global Hunger Index), as well as on the world's breadbaskets.

To achieve these goals, we use the Decision Support System for Agrotechnology Transfer (DSSAT) crop model to simulate changes in yields for rice, maize, and wheat following the adoption of different technologies, agricultural practices, improved varieties, or a combination of these, compared to a business-as-usual baseline. The results of DSSAT are then fed into IFPRI's IMPACT model (a partial equilibrium global agricultural sector model; see Chapter 3), using adoption pathways that consider profitability, initial costs and capital, risk-reduction, and complexity of the technology. IMPACT is then used to estimate global food supply and demand, food trade, and international food prices, as well as the resultant number of people at risk of food insecurity. In both models, the effects of the technologies are simulated under two alternative climate change scenarios.

Organization of the Book

The book is divided into six chapters. Chapter 2 describes the technologies evaluated in this study, providing the rationale for their selection and offering a detailed literature review to summarize the current knowledge regarding their effects on yields and on the use of resources, including water and energy inputs. Chapter 3 presents the modeling methodology in detail. Chapter 4 presents the main biophysical modeling results, and Chapter 5 discusses the economic modeling results. Chapter 6 discusses the policy implications of these results and offers conclusions.[3]

---

3 Appendixes that accompany this study can be found at http://www.ifpri.org/publication/food-security-world-natural-resource-scarcity.

Chapter 2

Technology Selection and Its Effects on Yields and Natural Resources

E xperts agree that increased production must be achieved by increasing
yields while using fewer resources and minimizing or reversing environmental impacts. This "sustainable intensification" approach is fundamentally about making the current agricultural system more efficient through the
use of new technologies[1] or by improving current production systems (Royal
Society 2009; Foley et al. 2011; Balmford, Green, and Phalan 2012; Garnett et
al. 2013; Smith 2013).

Sustainable intensification does not specify which agricultural technologies
and practices should be deployed, as these are context specific, but solutions
need to be environmentally sustainable (Garnett et al. 2013). Experts have
suggested that in many parts of the world, the adoption of small, incremental
changes—such as expanding fertilizer use, improving varieties, using mulches,
and using optimal spacing and precision agriculture in both high tech and low
tech systems—could have important positive effects on yields while limiting
environmental impacts (Royal Society 2009; Godfray et al. 2010; Clay 2011;
Foley et al. 2011; Balmford, Green, and Phalan 2012).

For this study, we selected both high- and low-tech solutions, ranging from
new traits in varieties (for example, drought-tolerant and heat-tolerant crops)
and water-saving irrigation technologies to practices that are considered more
efficient in terms of resource use (for example, integrated soil fertility management and no-till). Despite the current limitations on data availability, we also
included crop protection technology in the study, using estimates for chemical control to represent crop protection in general. The technologies assessed
were identified by experts from agricultural research organizations, the private
sector, and practitioners as key options to increase cereal yields rapidly and sustainably in the face of growing natural resource scarcity and climate change.
Once a preliminary set of technologies was identified, we used an online

---

1 The term "technology" refers to agricultural management practices, irrigation technologies, and
crop breeding strategies.

survey to solicit insights into the yield potential and natural resource impacts of these technologies. We also asked whether the selected technologies covered the spectrum of key technologies, and almost all experts who responded agreed that they did. A total of 419 experts responded to our survey, resulting in about 300 fully usable responses.[2]

The technologies cover a broad range of traditional, conventional, and advanced practices with some proven potential for yield improvement and wide geographic application. The chosen technologies are

1. no-till,

2. integrated soil fertility management (ISFM),

3. precision agriculture (PA),

4. organic agriculture (OA),

5. nitrogen-use efficiency (NUE),

6. water harvesting,

7. drip irrigation,

8. sprinkler irrigation,

9. improved varieties—drought-tolerant characters,

10. improved varieties—heat-tolerant characters, and

11. crop protection.

These technologies are at different stages of development and adoption across the world. Some are already in use in certain regions, whereas others are only at an exploratory phase. In agreement with the sustainable intensification strategy, the selected technologies and practices have the potential to increase yields while making better use of resources, helping farmers adapt to a changing climate, and reducing environmental impacts by limiting pollution and demands on ecosystem services. Specifically, many of these technologies have the potential to improve or restore soil fertility, thereby establishing conditions for increased productivity and higher resilience to drought conditions and climate variability (Molden 2007; Liniger et al. 2011) and therefore reducing production risk and encouraging additional investments in improved

---

2 The responses on the survey are available on request.

agricultural practices. These technologies are described in more detail in the remainder of this chapter.[3]

No-till

Although we focus here on no-till, under real farming conditions, the line between no-till and reduced till is frequently blurred, particularly in the case of smallholders, many of whom cannot implement no-till. No-till relies on three core activities:

- Absence of plowing with either broad castor direct seeding or placing the seeds in a shallow rut for protection from the elements or predators;

- Use of cover crops and mulching during part or all of the year;

- Crop rotation, in which the rotation often includes a main cash crop with one or more cover crops, to protect the soil surface for as long as possible.

No-till originated as a response to soil erosion, loss of soil organic matter, and consequent loss of soil fertility brought about by modern intensive agriculture in various parts of the world. In Brazil, the no-till revolution arose from widespread land degradation, which affected the south-tropical region of the country following the development of the Cerrados in the 1970s and translated into loss of soil organic matter, soil compaction, reduction in water infiltration, and pollution of waterways through erosion and runoff (Bollinger et al. 2006). Worldwide no-till increased from 45 million hectares in 2001 to more than 100 million hectares in 2008 (Derpsch and Friedrich 2009). In 2007, 26 percent of total cropland in the United States was under no-till, compared with 45 percent in Brazil,[4] 46 percent in Canada, 50 percent in Australia, 69 percent in Argentina, and up to 80 percent and 90 percent in Uruguay and Paraguay, respectively (Bollinger et al. 2006; Derpsch and Friedrich 2009).

The span of no-till from regions close to the Arctic Circle (for example, Finland) to the tropics (for example, Kenya and Uganda) and from sea level to high altitudes (for example, Bolivia) shows its adaptability and economic viability under different cropping systems as well as different climatic and soil conditions (Table 2.1).

---

3 Heat tolerance and improved nitrogen-use efficiency are still in the exploratory stage of development. We therefore include only brief descriptions of these two technologies in this literature review.

4 Bollinger et al. (2006) report that this percentage may be up to 80 percent in southern Brazil.

TABLE 2.1   Area under no-till, by continent

Continent	Area (thousand ha)	Share of total (%)
South America	49,579	46.8
North America	40,074	37.8
Australia and New Zealand	12,162	11.5
Asia	2,530	2.3
Europe	1,150	1.1
Africa	368	0.3
World	105,863	100.0

Source: Derpsch and Friedrich (2009).
Note: Total area under no-till in the Indo-Gangetic Plain of South Asia was estimated at 1.9 million hectares in 2005. Derpsch and Friedrich (2009) did not include the Indo-Gangetic Plain in their estimates, because the soil is tilled to prepare it for rice in this rice-wheat system of double cropping.

Most adoption is taking place on medium to large farms; adoption by smallholder farmers appears to be less common, with the exception of Brazil (Bollinger et al. 2006; Derpsch and Friedrich 2009). The New Partnership for Africa Development and the Alliance for Green Revolution in Africa have incorporated no-till in regional agricultural policies, and in southern and eastern Africa, the number of farmers adopting no-till has reached 100,000 (Derpsch and Friedrich 2009).

The literature offers many studies on the effects of no-till on yields and the use of resources under different cropping systems. No-till promotes soil fertility by improving both soil structure and soil organic carbon content; residues and cover crops induce accumulation of organic matter (at least in the surface soil horizon), conserve humidity, and protect the soil from water and wind erosion (Hobbs, Sayre, and Gupta 2008).

Conventional tillage loosens and aerates the soil, increasing microbial oxidation of organic matter to CO_2 (Hobbs, Sayre, and Gupta 2008; Giller et al. 2009; Kassam et al. 2009; de Rouw et al. 2010). In contrast, no-till increases soil organic matter, which supports the role of agriculture in carbon sequestration and mitigation of climate change. The soils that are the most vulnerable to tillage-induced loss of organic matter are coarse-textured soils and those with low-activity clays of the tropics and subtropics.

Studies have also shown that no-till enhances water-use efficiency, mainly by reducing runoff and evaporative losses and by improving water infiltration (Hobbs, Sayre, and Gupta 2008). Hobbs, Sayre, and Gupta (2008) and Kassam et al. (2009) report that yields under no-till can be equal to or higher than yields under conventional tillage, and that the essential improvement brought about by no-till consists of greater yield stability over time. Other studies found increasing yields for wheat (by 5–7 percent) in the Indo-Gangetic plains (Erenstein 2009), and for maize (30 percent) in the highlands of central Mexico, in combination with rotation of crops and use of residues as soil cover (Govaerts, Sayer, and Deckers 2005). No-till gave higher yields for wheat, maize, and teff in Ethiopia, and for maize in Malawi and Mozambique on smallholder plots ranging from 0.1 to 0.5 hectares (Ito, Matsumoto, and Quinones 2007).

It is difficult to incorporate fertilizers into soils with low infiltration rates, so that using no-till on them may result in higher nutrient losses in runoff (Lerch et al. 2005). In the first years of using no-till, residues on the soil surface may immobilize nitrogen in the topsoil, so that more fertilizer may be needed to compensate (Bollinger et al. 2006). Moreover, residues are no longer mixed with the soil, which may slow mineralization, induce faster denitrification and leaching, and increase volatilization (Cantero-Martinez, Angas, and Lampurlanes 2003). The effect is greater for heavier-textured soils.

Energy requirements appear to be lower for no-till compared to conventional systems. Mrabet (2008) found that for large producers, conventional tillage can use more than three times as much fuel and tends to require higher machinery costs compared to no-till. Other studies similarly suggest that no-till is associated with lower fuel requirements than conventional tillage, because it uses smaller tractors and because fewer passes are needed with the tractor (FAO 2001; Pieri et al. 2002). Zentner et al. (2004) determined that no-till can enhance the use efficiency of nonrenewable energy sources when adopted in combination with diversified crop rotations.

Adoption of no-till is affected by a range of often context-specific factors. The availability of herbicides, particularly glyphosate, has been cited as the single most important factor encouraging the successful spread of no-till in Brazil (Bollinger et al. 2006), and the availability of glyphosate-resistant crops was critical for the expansion of no-till in the United States (Givens et al. 2009). The cost of inputs may significantly influence the profitability of a farm, and as a result, this technology may not be ideal for smallholder farmers. In SSA, where smallholders often practice a mixed agriculture-livestock system, residues from cropping are a precious source of fodder, and the scarcity of material caused by dry conditions does not always allow smallholders to spare biomass for mulching. Therefore, in this region the availability of mulch for cover and nutrients can be a critical constraint to adoption of no-till (Giller et al. 2009).

There is general agreement that no-till reduces labor requirements and can reduce production costs. The elimination of plowing allows for cost control through reduction of labor and fuel needs (Bollinger et al. 2006; Dumanski et al. 2006; Derpsch and Friedrich 2009; Kassam et al. 2009). A study in the Indo-Gangetic plains showed that, when including savings in costs of production, no-till brought about an increase in farm income from wheat production of US$97/hectare (an increase in real household incomes of US$180–340 per farm) (Erenstein 2009). In China, the adoption of no-till for wheat production raised yields and reduced production costs, hence causing an increase of 30 percent in net average economic returns over 4 years (Du et al. 2000; Wang et al. 2009).

A no-till system requires herbicides to substitute for tillage in controlling weeds (FAO 2001). As herbicides are petroleum-based products, an increase in crude oil prices would increase their cost and could partially or completely offset the advantage obtained through lower fuel usage. However, a study by Sanchez-Giron et al. (2007) in Spain showed that even considering the higher herbicide costs per hectare, total economic performance in terms of profit and net margin (in euros/hectare/year) was consistently higher for no-till, regardles of the size of the farm.

Overall, higher fuel prices should favor the expansion of conservation agriculture (minimum tillage as well as no-till). A study in the United States shows a significant—but small—positive effect of the price of crude oil on the expansion of conservation agriculture: a 10 percent increase in the price of oil triggered an expansion of area under conservation agriculture by 0.4 percent (FAO 2001). Interestingly, the expansion did not involve the adoption of conservation agriculture by new users and was instead due to the expansion of area under conservation agriculture by users that had already adopted it on part of their land (FAO 2001).

Integrated Soil Fertility Management

The goal of integrated soil fertility management (ISFM) is to increase productivity by ensuring that crops have an adequate and balanced supply of nutrients (Gruhn, Goletti, and Yudelman 2000) and maximizing their efficient use. ISFM seeks to maximize agronomic efficiency by combining a balanced nutrient supply with improved varieties and agronomy adapted to local conditions (Vanlauwe et al. 2011). Synthetic fertilizers and organic inputs bring different benefits to the soil. Both are sources of nutrients, but livestock manures, crop residues, and compost also increase the soil organic matter, which improves soil structure and nutrient cycling and increases soil health and fertility (Mateete, Nteranya, and Woomer 2010).

Although organic matter is particularly important in SSA, the profitability of using organic material can change significantly based on the distance to market and transportation method. Therefore, an incentive exists to produce organic inputs in situ, but here the practice is encountering land and labor constraints or growing opportunity costs. This is particularly true as plots of land in SSA are becoming smaller, making it more difficult for smallholders to produce sufficient amounts of organic nutrient sources (Place et al. 2003).

Vanlauwe et al. (2011) and Chivenge, Vanlauwe, and Six (2011) conclude that the combination of fertilizer and organic inputs leads to higher yields compared to a control with no fertilizers and compared to a control with only chemical fertilizers or only organic inputs. Chivenge, Vanlauwe, and Six (2011) show that yield responses increased with increasing quality of organic input and also with increasing quantity of organic-nitrogen. Moreover, organic material, alone or in combination with chemical nitrogen, led to more accumulation of soil organic carbon compared to a control without nutrient inputs, or a control with only chemical nitrogen inputs. The authors also find that the effects for yields and soil organic carbon were stronger in sandy soils compared to clayey or loamy soils.

A survey study conducted in nine villages in Kirege, Kenya, investigated the factors affecting smallholder decisions on ISFM adoption. The study shows significant correlation between perception of soil fertility as a current problem and adoption of ISFM technology; hence, sensitizing farmers about their soil fertility status may promote adoption (Mugwe et al. 2009). The number of months during which households had to buy food to close the food deficit was also a major factor, along with the ability to hire labor on a seasonal basis, as the ISFM technology is labor intensive.

Precision Agriculture

Precision agriculture (PA) is "a way to apply the right treatment in the right place at the right time" (Gebbers and Adamchuck 2010, 828) by optimizing the use of available resources (such as water, fertilizer, or pesticides) to increase production and profits. PA, which started in the mid-1980s, came from understanding the mechanisms that link biophysical conditions to variability in crop yields. Developments in information and automation technologies allowed variations in crop yield to be quantified and mapped, and hence the biophysical determinants to be managed precisely (Bramley 2009; Gebbers and Adamchuck 2010).

PA is based on a set of data-gathering technologies, ranging from on-the-ground sensors and satellite imagery to the Global Positioning System (GPS) and geographic information systems, which provide high-resolution bio-physical and crop-related data (Bramley 2009). Variable rate technology[5] is the most widely practiced PA method. It relies on data from soil sampling, yield monitors, and remote or proximal sensing to create yield maps and regulate the amount and timing of application of water and agro-chemicals, especially nitrogen (Gebbers and Adamchuck 2010). Yield monitors are the single most common PA technology used around the world; 90 percent of adopted yield monitors are in the United States, followed by Germany, Argentina, and Australia (Griffin and Lowenberg-DeBoer 2005).

Studies of the effects of PA on crop yields are rare, and the few published studies show mixed results (for example, see Ferguson et al. 1999 as cited in Cassman 1999). In general, different sections of a putatively uniform field have substantially lower yield potential than the median value of the whole field. The objective of PA is to apply less fertilizer to these lower-yielding microsites and apply more to those sites with higher yield potential (instead of applying fertilizer uniformly across the whole field). This strategy can increase the total yield of the field, because fertilizer is applied to those microsites that can respond better. However, whether the yield of the whole field increases depends on how the crops respond to the nutrient (that is, on the yield response curve) and on the soil type.

Bongiovanni and Lowenberg-DeBoer (2004) conclude that PA can benefit the environment, as the more targeted use of inputs (both nutrients and herbicides) reduces losses from excess applications. Some energy savings have been reported, mainly resulting from lower nutrient use (Lowenberg-DeBoer and Griffin 2006), and site-specific nutrient applications are reported to reduce nitrate leaching and to increase nitrogen-use efficiency (NUE) (Cassman 1999). However, application of variable rate technology does not necessarily mean that the application of inputs like nitrogen will be lower (Harmel et al. 2004), as this depends on the share of areas in a field with high potential (and thus higher nitrogen application levels). An example from the sugarcane and dairy industry in Australia shows that NUE can be improved through yield mapping, resulting in benefits for water quality (Bramley et al. 2008).

In terms of economic benefits, some PA tools are labor saving (for example, GPS guidance) (Lowenberg-DeBoer and Griffin 2006), but managerial time is high, at least during the early stages of adoption (Daberkow and McBride 2003). In a review of 234 studies published from 1988 to 2005 (Griffin and Lowenberg-DeBoer 2005), PA was found to be profitable in 68 percent of the

---

5   That is, the use of sensors and other technologies for targeted application of inputs.

cases. Most studies were done on maize (37 percent) or wheat (11 percent). Of these, 73 and 52 percent reported benefits, respectively.

Silva et al. (2007) analyze the economic feasibility of PA (yield maps and soil mapping) for maize and soybeans in the state of Mato Grosso do Sul, Brazil, compared with conventional farming. The authors find that, on average, PA is more costly than conventional farming for both crops, mainly because of the need for qualified labor, technical assistance, maintenance of equipment, yield maps, and soil mapping. However, PA led to higher yields and higher gross revenue.

PA has not been widely adopted by farmers (Fountas, Pedersen, and Blackmore 2005), and as of 2001, most adopters were in Australia, Canada, the United States, Argentina, and Europe (Swinton and Lowenberg-DeBoer 2001). A suite of socioeconomic, agronomic, and technological challenges limit the broader adoption of PA (Robert 2002). Lack of basic information, absence of site-specific fertilizer recommendations, and lack of qualified agronomic services compound multiple technological barriers related to the availability and cost of the technology, such as machinery, sensors, GPS, software, and remote sensing (Robert 2002). McBratney, Whelan, and Ancev (2005) derived indicators of a country's suitability for adopting PA and estimated that countries with large cropland area per farm worker (as well as large fertilizer use per hectare) are likely to benefit best from PA methods.

Organic Agriculture

Organic agriculture (OA) is regulated in its definition, guiding principles, and implementation by several international associations (Gomiero, Pimentel, and Paoletti 2011). OA excludes the use of most synthetic inorganic fertilizers, chemical pest controls, and genetically modified cultivars. OA promotes a range of agronomic interventions to increase soil fertility and relies on biological processes to control emergence of weeds and pests (Hendrix 2007; Connor 2008; Seufert, Ramankutty, and Foley 2012).

A global assessment conducted by Badgely and colleagues concluded that organic agriculture could achieve yields similar to or greater than conventional agriculture, therefore having the potential to contribute substantially to global food supply (Badgley et al. 2007). They further argued that legumes used as green manure could provide "enough biologically fixed nitrogen to replace the entire amount of synthetic nitrogen fertilizer currently in use" (Badgley et al. 2007 [for quote, see abstract]; Badgley and Perfecto 2007). The conclusions of this study have been disputed on several grounds (Cassman 2007;

Hendrix 2007; Connor 2008). Re-examination of the published papers on which Badgley et al. (2007) based their argument shows that when yields from OA crops equaled or exceeded those of conventionally farmed crops, they had received similar amount of nitrogen in the organic material applied, much of which came from outside the system (Kirchmann, Kaetterer, and Bergstroem 2008). Therefore, OA can make a substantial contribution to the global food supply only at the cost of expanding the global cropped area; the same conclusion applies to using legumes to substitute for nitrogen fertilizer.

Two recent metastudies showed that yields from OA average 20–25 percent less than those from conventional agriculture, but with large variations (de Ponti, Rijk, and Ittersum 2012; Seufert, Ramankutty, and Foley 2012). Seufert, Ramankutty, and Foley (2012) show that although yields of organic fruit and oilseed are only 3 and 11 percent less, respectively, than those of conventional agriculture, yields of organic cereals and vegetables are 26 and 33 percent less, respectively.

In terms of natural resource use, Pimentel et al. (2005) and Tuomisto et al. (2012) report that OA systems require between 21 percent and 32 percent less energy compared to conventional systems. Reliance on manure and organic inputs leads to more stable soil aggregates and therefore reduced erosion. Soil losses under OA were less than 75 percent of the maximum loss-tolerance in the region, whereas with conventional agriculture, the loss was three times the maximum loss-tolerance (Reganold, Elliott, and Unger 1987).

By increasing soil organic matter content, OA improves soil structure and increases the water-holding capacity of the soil and is therefore more tolerant of drought (Pimentel et al. 2005). Nitrogen leaching and emissions of nitrous oxide and ammonia per unit area are lower in OA compared to conventional agriculture because of the lower nitrogen inputs, but they are larger per unit of product because of OA's lower yields (Pimentel et al. 2005; Balmford, Green, and Phalan 2012; Tuomisto et al. 2012).

OA increases soil microfauna populations and microbial biomass, and it promotes higher species abundance compared to conventional agriculture (Pimentel et al. 2005; Tuomisto et al. 2012). In small-scale agricultural landscapes with a variety of biotypes, however, OA does not increase species abundance compared with conventional agriculture (Gomiero, Pimentel, and Paoletti 2011).

In terms of economic profitability, Hendrix (2007) reports that costs to protect soil fertility on organic maize farms is 40 percent higher than on conventional farms, and costs are driven up by pest pressure, as yields are limited to 80–85 percent of the yields of conventional farms. Pimentel et al. (2005) report that organic systems may need between 15 and 75 percent more labor inputs compared to conventional systems, and when including the costs of family labor and those of the initial transition to organic, the average net returns per hectare for OA were 22 percent lower than for conventional agriculture.

OA is currently practiced on only 37 million hectares, or less than 1 percent of the global agricultural area, with most of the production concentrated in developed countries (Willer and Kilcher 2011).

Nitrogen-Use Efficiency (NUE)

The ability of a plant to absorb and use the available nitrogen depends on many variables, including the competing use of nitrogen by soil microorganisms and losses through leaching (Pathak, Lochab, and Raghuram 2011). Roberts (2008, 177) defines agronomic NUE as "nutrients recovered within the entire soil-crop-root system" and recognizes that in the context of food security, the efficiency of use of nutrients has to be optimized in a system that strives to increase yields and achieve economic viability (Dibb 2000; Roberts 2008). However, several common definitions of NUE exist,[6] and the appropriate adoption of one definition or the other is dependent on the crop and the physiological processes involved in the efficient uptake and use of nitrogen (Pathak, Lochab, and Raghuram 2011). When expressed as yield of grain per unit of nitrogen in the soil (both from residues and fertilizers), NUE in cereals is estimated to be below 50 percent. Therefore, significant opportunities still exist for improving NUE in cereals through a combination of changes in agricultural management practices (for example, improving the synchrony between the crop demand and supply of nitrogen) and by identifying and selecting new hybrids and genetic markers (or both) for molecular breeding (Hirel et al. 2007; Pathak, Lochab, and Raghuram 2011). No transgenic or classically bred NUE-improved crops have yet been released for commercial use, yet promising advances are being made in the field through the conventional or molecular marker-assisted breeding to enhance the plants' innate physiological ability to uptake or assimilate nitrogen (Pathak, Lochab, and Raghuram 2011).

---

6 A few common agronomic indices used to describe NUE are

1. partial factor productivity (kilogram of crop yield per kilogram of nutrient applied, or the ratio of yield to the amount of applied nitrogen) (Dobermann and Cassman 2005),
2. agronomic efficiency (kilogram of crop yield increase per kilogram of nutrient applied),
3. apparent recovery efficiency (kilogram of nutrient taken up per kilogram of nutrient applied),
4. physiological efficiency (kilogram of yield increase per kilogram of nutrient taken up), and
5. crop removal efficiency (removal of nutrient in harvested crop as a percentage of nutrient applied).

Water Harvesting

Two categories of harvesting rainwater are recognized (Ngigi 2003):

1. In situ water harvesting: Crop and soil management that captures rainwater and stores it in the root zone of the soil profile for subsequent root uptake. In situ systems include tillage practices, residue management, and management of soil fertility; they typically conserve water in the soil profile for a few days to weeks.

2. Runoff harvesting: Plant water availability is maximized by harvesting surface runoff for supplemental irrigation of the same crop for storage to be used on subsequent crops.

Because of the costs of construction and implementation, most water harvesting practices in arid and semi-arid environments consist of either in situ or direct application of runoff. However, the use of storage systems is increasing (Rockström, Barron, and Fox 2002).

Water harvesting has been practiced for centuries in the Middle East, North Africa, SSA, Mexico, South Asia, and China (Critchley and Siegert 1991; Ngigi et al. 2005; Oweis and Hachum 2009). Although adoption is widespread, adoption levels in any given region or country remain low.

Water harvesting increases crop yields. In China's semi-arid Gansu Province, supplementary irrigation by harvested water increases yields of intercropped maize by 90 percent and of wheat by 63 percent, compared with rainfed crops (Yuan, Li, and Liu 2003). Irrigation with rainwater harvested from a macro-catchment in the Makanya River watershed in Tanzania in 2004 gave yields in the short rainy season that were almost double the national and regional averages (Hatibu et al. 2006). Similarly, in microcatchments in the Mwanga district of Tanzania, water harvesting more than doubled yields of maize in the short rainy season (Kayombo, Hatibu, and Mahoo 2004).

Water harvesting appears to increase biodiversity at the field and landscape levels by recharging aquifers, which stimulates regrowth of vegetation and greater diversity of plant species (Vohland and Barry 2009). In turn, increased availability of biomass for food and shelter often correlates with greater abundance of animal species and more complex trophic chains. However, rainwater harvesting is often used to cultivate crops that replace indigenous grasses and herbs, so the overall outcome is uncertain.

Water harvesting upstream may reduce the amount of water available downstream (Ngigi 2003; Wisser et al. 2010). In the Volta Basin, several thousand small reservoirs have been constructed for domestic and stock water and small-scale irrigation. When assessing whether they would impact on downstream water flow, Lemoalle and de Condappa (2012, 210) write, "Very strong development of small reservoirs (up to seven times the present number) would only decrease the inflow to Lake Volta . . . by 3% in the present climatic conditions."

In terms of economic efficiency, water harvesting generally increases profits, but it is often difficult to determine labor costs adequately for the structures (Isika, Mutiso, and Muyanga 2002; Fox, Rockström, and Barron 2005; Hatibu et al. 2006).

Drip Irrigation

Drip irrigation is a system of water delivery for agricultural crops that releases minute quantities of water directly onto the root zone of the plant (Goldberg, Gornat, and Rimon 1976) using tubes and emitters that distribute the water and sometimes using soluble fertilizer as well (Burney and Naylor 2012). Depending on the context, there can be wide variations in the implementation. In developed countries, emitters are often pressure regulated to enable one pump to irrigate large areas (Burney and Naylor 2012). In developing countries, the systems are often smaller, simpler, and cheaper, using drip lines fed from small raised tanks (Upadhyay, Samad, and Giordano 2005; Burney and Naylor 2012).

Drip irrigation was developed in Israel to deal with water scarcity. It is used in countries on all continents, but in many, the rates of adoption are low. India and China have the largest areas under drip irrigation, followed by the United States, Spain, Italy, Korea, South Africa, Brazil, Iran, and Australia (ICID 2012). But in many of these countries, drip irrigation makes up only a small fraction of the total irrigation. In terms of the fraction of total irrigated land using drip irrigation, Israel ranks first (73.6 percent), followed by Estonia (50 percent), Spain (47.8 percent), Korea (39.6), South Africa (21.9), Italy (21.3), Finland (14.3), Saudi Arabia (12.2), Slovenia (9.6), and Malawi (9.1). (Calculated from data in ICID 2012.)

The advantage of drip irrigation is that farmers can control the timing and amount of irrigation, which both increases the yield and improves the quality of the product (Cornish 1998). Slow distribution of water over the growing season means that plants should not suffer water stress and can produce consistently high yields (IDE, n.d.; Möller and Weatherhead 2007). Commercial cotton farms in India produced yield increases of 114 percent under drip irrigation by avoiding water stress, supplying water directly to the root zone so that none was wasted, and increasing nutrient uptake by delivering fertilizer to the roots (Narayanamoorthy 2008). However, a recent review of drip irrigation adopters' experiences in four SSA countries found that fewer than half cited an increase in productivity or yield as a benefit (Friedlander, Tal, and Lazarovitch 2013).

In terms of resource use, efficiency of water use is an important benefit of drip irrigation, with water savings of 20–80 percent compared with furrow or flood irrigation (Sivanappan 1994; Hutmacher et al. 2001; Alam et al. 2002; Godoy et al. 2003; Maisiri et al. 2005). Furthermore, drip irrigation loses little water through conveyance (INCID 1994; Narayanamoorthy 1996, 1997; Dhawan 2000), resulting in irrigation efficiencies[7] of more than 90 percent (Cornish 1998). These efficiencies could be further increased by controlling water application to prevent water percolation below the root zone (Bergez et al. 2002; El-Hendawy, Hokam, and Schmidhalter 2008).

Drip irrigation reduces the labor needed for irrigation, fertilizing, and weeding (Cornish 1998; IDE, n.d.), with farmers often identifying labor savings as the main factor driving the adoption of this technology (see the review in van der Kooij et al. 2013). Drip irrigation can reduce labor requirements by 50 percent, although these savings apply mainly to larger-scale commercial operations (Dhawan 2000). Drip kits for small fields did not increase labor savings compared with applying water directly to the field (Kabutha, Blank, and Van Koppen 2000; ITC 2003; Moyo et al. 2006), although a review of drip irrigation in Nepal found that in women's home vegetable plots, drip irrigation reduced the labor required for irrigation by 50 percent (Upadhyay, Samad, and Giordano 2005).

Commercial drip irrigation on a tea plantation in Tanzania required that yield increase by 410 kilograms/hectare to offset the investment and higher management costs (Moller and Weatherhead 2007). Low-cost drip irrigation for the poorest in Nepal was profitable with a relatively high internal rate of return on the investment (Upadhyay, Samad, and Giordano 2005).

Sprinkler Irrigation

Sprinkler irrigation is a method of applying water to crops that mimics rainfall and aims at distributing water uniformly across the field to promote better crop growth (Brouwer et al. 1988). Water is distributed under pressure

---

7 Irrigation efficiency is defined as the proportion of water used (that is, applied to the field or crop) that is actually consumed by the crop (Perry et al. 2009).

through a system of pipes and is sprayed onto the crop using nozzles. Sprinkler irrigation is suitable for a variety of row and field crops, and it can be adapted to different slopes and farming conditions (Brouwer et al. 1988). Similar to drip irrigation, sprinkler systems allow distribution of precise amounts of water following a predetermined schedule, thereby enabling a more efficient use of water. This practice is especially beneficial as an adaptation to climate change and in areas where water supply is irregular and unreliable. In these areas and conditions, the improved efficiency of water use can help increase crop yields (Lecina et al. 2010). Sprinkler systems are available for both small- and large-scale applications. The size of the farm and especially the availability of capital, labor, and energy (for example, engines and electricity) determine the choice of the system (for example, one that is hand operated or mechanically operated).

Estimates of the extent of adoption of sprinkler irrigation systems vary substantially. Kulkarni, Reinders, and Ligetvari (2006) placed the adoption at 13.3 million hectares in the Americas, 10.1 million hectares in Europe, 6.8 million hectares in Asia, 1.9 million hectares in Africa, and 0.9 million hectares in Oceania. Data from AQUASTAT[8] (the water information systems of the Food and Agriculture Organization of the United Nations) shows the largest adoption in a region made up of Eastern Europe and central Asia, followed by Western Europe.

Most commonly, the drive behind the adoption of modern irrigation technologies is the need to achieve better irrigation efficiencies and water savings in response to declining water supply following population growth, economic development, climatic changes, or a combination of these factors (Kahlown et al. 2007; Lecina et al. 2010; Zou et al. 2013). However, the factors that drive the adoption of sprinkler or drip irrigation are many and differ from region to region. In Spain, the modernization of irrigation infrastructure was driven mostly by the liberalization of agricultural markets and the falling availability of agricultural labor, which pushed farmers toward a more flexible system of production (Lecina et al. 2010).

In South Asia as in other parts of the world, the agriculture sector is being pressured to reduce water consumption and make it available for the urban and industrial sectors. Adoption of sprinklers in India, across different topography and climatic conditions, has improved irrigation efficiencies by up to 80 percent (Sharma 1984). Kahlown et al. (2007) tested the potential of rain-gun sprinklers to improve the irrigation efficiency and therefore the water

---

8 http://www.fao.org/nr/water/aquastat/dbase/index.stm, accessed September 2013.

productivity[9] of rice and wheat cultivation in the Indo-Gangetic plains of Pakistan. They found that the use of sprinklers increased yields and crop water productivity compared to traditional irrigation. However, in Pakistan as elsewhere, the potential for adoption of sprinklers to irrigate rice and wheat is affected by cost-benefit considerations, especially the value of water saved and potential yield increases versus expenses for on-farm water storage, as well as for the purchase and maintenance of the sprinkler system. At 2007 market costs and prices (of water and crops), the use of sprinkler irrigation was a financially viable solution in Pakistan. Water productivity increases would have resulted in net benefits, even considering all the costs associated with sprinkler irrigation: capital and maintenance costs, as well as those for the pumps and for the on-farm water storage (Kahlown et al. 2007).

Modern irrigation systems like sprinklers (or drip irrigation) have the potential to maximize transpiration and minimize evaporation, that is, divert nonbeneficial water consumption to beneficial consumption. Several studies show that although the application of irrigation water through sprinklers can result in larger biomass production and increases in crop yields at the single farm or plot scale, it might not translate into the desired water savings at the basin scale (Ward and Pulido-Velazquez 2008; Perry et al. 2009; Lecina et al. 2010) as water use patterns change (Lecina et al. 2010). Farmers also increase cropping intensity, because the pressurized systems used in sprinkler irrigation have higher conveyance capacity. Because of the greater cropping intensity, better irrigation application efficiency, and wind and evaporative losses, sprinkler-irrigated areas benefit from higher yields and higher production levels, but both consumed and depleted water fractions are larger compared to surface irrigation (Lecina et al. 2010).

The two main constraints on the adoption of sprinkler irrigation are (1) the cost and knowledge requirements of the system itself and (2) the need for labor to install, move, and maintain pipes and sprinklers around the fields (Brouwer et al. 1988). As the primary goal of sprinkler irrigation is to provide uniform irrigated conditions to the root zone, several sprinklers usually must be placed in close proximity to one another (Brouwer et al. 1988). Costs and availability of labor are an additional concern, especially for smallholders. Because of these constraints, sprinklers are often adopted by farmer groups or cooperatives to share the high fixed costs and the burden of installation, management, and maintenance.

---

9 Water productivity is defined as the ratio between the amount of crop produced and the amount of water consumed to obtain such production (Perry et al. 2009).

Improved Varieties—Drought-Tolerant Characters

Maize

As a C4 plant, maize has some inherent advantages under drought conditions (Lopes et al. 2011); however, drought is the main constraint to maize yields in both temperate and tropical regions, and it is one of the causes for the difference in average productivity between them. Edmeades (2008) reports that as most maize is globally grown in rainfed conditions, average annual yield losses stemming from drought are 15 percent globally. These losses are greater in tropical countries, where maize production is affected by high rainfall variability (Edmeades 2008).

Barnabas, Jager, and Feher (2008) describe drought resistance for maize arising from three different possible strategies:

1. Escape: Successful reproduction before onset of severe stress by means of short crop duration, high growth rate, efficient storage, and use of reserves for seed production;

2. Avoidance: Maintenance of high tissue water status during stress periods (by minimizing water loss through stomatal closure, reduced leaf area, and senescence of older leaves), or maximizing water uptake (by increasing root growth and modifying crop architecture); and

3. Tolerance per se: Physiological and cellular adjustments to tolerate tissue water desiccation (these are internal osmotic adjustments or other structural changes that allow the plant to function under water stress and to recover function after the stress is relieved).

Adaptation to abiotic stress is a trait controlled by many genes. Breeding targeted to protect yields in drought-prone climates has to focus on changes at flowering or during early grain development, because maize is most sensitive to drought during these stages (Lopes et al. 2011). Flowering is critical because the male and female flowers are physically separated on the maize plant, and they respond differently to water deficits, which can cause asynchrony in their flowering times. Asynchrony can thwart or reduce fertilization, reducing grain filling and yields (Grant et al. 1989; Cairns et al. 2012).

Secondary traits are postulated to increase drought resistance (Bruce, Edmeades, and Barker 2002; Barnabas, Jager, and Feher 2008; Edmeades 2008; Lopes et al. 2011; Messmer et al. 2011):

1. High level of synchrony of male and female flowering, so that they occur simultaneously as near as possible;

2. Reduced plant density (as implemented by farmers in the sub-Sahel);

3. Changes in carbon allocation pattern to build deep root systems before the onset of drought (although deep root systems only confer advantage in deep soils, not in shallow ones);

4. Higher root biomass and improved root architecture to increase the crop's ability to take up water;

5. Leaf curling (or rolling) to reduce transpiration without much reduction of leaf photosynthesis (the canopy structure of maize and other C4 monocotyledonous plants allows leaf curling); and

6. Increased stay-green (low rates of leaf senescence favors grain fill under drought), but the stay-green must be functional.

Breeding strategies will target one or maybe several of these secondary traits depending on the drought scenario in question. Stay-green allows maize to maintain its vegetative biomass, so that it can contribute to yield under mild to moderate water deficit (Lopes et al. 2011). Under severe water deficits, the strategy is to reduce the risk of crop failure, with low but stable yields, which is a strategy that forgoes high yields in good years. This is in line with an escape strategy, which shortens the life cycle, and with traits that lead to water conservation like reduction in leaf area, low stomatal conductance, high water-use efficiency and "deep but sparse root system[s]" (Lopes et al. 2011, 3138).

Maize, which is a C4 plant, can perform better in drought compared to C3 plants (Lopes et al. 2011). However, drought still constrains maize yields throughout its geographic range. Maize production in southern Africa was only 12.5 million tons in 1992, a year of drought, compared with 23.5 million tons in 1993 (Bänziger and Araus 2007).

In the past few years, the Drought-Tolerant Maize for Africa project has facilitated the release in several African countries of 53 drought-tolerant varieties, both hybrids and open pollinated varieties, based on International Maize and Wheat Improvement Center (CIMMYT) and International Institute of Tropical Agriculture germplasm (Prasanna et al. 2011). "DT [Drought-tolerant] maize currently occupies approximately 2 million hectares (mha) in Africa, yielding at least 1 t/ha [metric ton/hectare] more than the local varieties under drought stress conditions" (Prasanna et al. 2011, 5). The most promising variety, ZM521, is sown on more than 1 million hectares in southeast Africa (Edmeades 2008).

The private sector has also registered some success in improving drought-tolerant hybrids thanks to multi-environment trials and to molecular breeding. The adoption of marker-aided selection has "virtually doubled the rate of genetic gain in Monsanto's maize population" (Edmeades 2008, 206). In 2010 the Swiss agribusiness Syngenta presented a drought-tolerant maize strain with the declared potential to increase yields by 15 percent in water-stressed environments (Tollefson 2011). The following year Pioneer Hi-Bred International announced drought-tolerant maize hybrids, with the potential for a 5 percent yield increase in field trials, which will soon be marketed in the United States (Tollefson 2011).

Maize is the most advanced of the drought-tolerant crops under biotech development. "The first biotech maize hybrids with a degree of drought tolerance are expected to be commercialized by 2012 in the USA, and the first tropical drought tolerant biotech maize is expected by 2017 for Sub Saharan Africa" (James 2010, 10). Preliminary projections for the United States indicate that yield gains from genetically modified drought-tolerant maize could be between 8 and 10 percent in the non-irrigated areas (from North Dakota to Texas). It is also projected that yields in the dry regions may increase from 5.5 to 7.5 metric tons[10] per hectare by 2015 (James 2009).

Rice

Among cereals, the rice plant is the most sensitive to water stress, having evolved in waterlogged environments; drought is the main global constraint to rice yields (Bouman et al. 2007). Growing competition for water resources as well as changing weather and rainfall patterns are particularly affecting rainfed environments but also water-constrained irrigated areas that depend on surface water for irrigation (Serraj et al. 2011).

From the point of view of genetic improvement, developing drought-resistant rice varieties has been complicated by the difficulty of screening for the key traits, and progress has been slow. Researchers and farmers are looking for traits of drought tolerance accompanied by high-yielding potential both under drought-stressed and unstressed conditions. This requirement is key for varieties that must be adapted to unpredictable rainfall; achieving this goal would lower production risk and encourage farmers to invest in agricultural inputs and other yield-enhancing practices (Verulkar et al. 2010).

In 2011 the Nepalese Institute for Agriculture and Animal Science released three rice varieties suitable for the drought-prone areas of the western mid-hills of Nepal, developed by the International Rice Research Institute (IRRI) through the project Stress-Tolerant Rice for Africa and South Asia (Kumar and Frio

---

10   In this book, all tons refer to metric tons.

2011).[11] Between 2008 and 2010 the released varieties, dubbed Sukha-1, Sukha-2, and Sukha-3, were tested for yields under drought. They were chosen not only for their drought tolerance but also for other characters popular among farmers, including an ability to be grown both as upland rice and as lowland rainfed rice, early maturity, high grain yield, improved milling recovery, tolerance to diseases, and easy threshing (Kumar and Frio 2011).[12]

One of the *Sukha-dhan* varieties has been successfully tested for use during the *monga* season[13] in drought-prone areas of Bangladesh, where it maintained yields of 4.0–4.5 metric tons/hectare (Neogi and Baltazar 2011). It has been released for commercial cultivation as BRRI *dhan* 56 (Kumar and Frio 2011). BRRI *dhan* 56 and BRRI *dhan* 57 (another variety released in Bangladesh) are not only drought tolerant, but they also allow farmers to escape late-season drought thanks to their rapid maturity (about 100 days required before harvest) (Kumar 2011). Farmers found that because of the early maturity and medium-sized grain, the drought-tolerant varieties can command a higher price in the market, increasing the profitability of the harvest. A one-year study found that farmers could have a net return of 19,200 Bangladeshi taka/hectare (about US$230/hectare) (Neogi and Baltazar 2011).

In 2010 Ghaiya 1, another variety developed at IRRI, was released for rainfed upland systems, which cover one-tenth of all rice-cultivated areas in Bangladesh (between altitudes of 300 and 750 meters) and experience erratic rainfall and drought stress.[14]

Wheat

About 50 percent of the global area devoted to wheat production is affected by drought (Pfeiffer et al. 2005). As for other cereals, drought-induced yield damage is more likely when drought occurs during flowering and grain filling.

Most wheat-breeding efforts by CIMMYT focus on the common spring bread wheat, which covers about 95 percent of world production (Ortiz et al. 2008). As for maize and rice, researchers aim at producing improved wheat

---

11 IRRI started the project in 2007 in collaboration with AfricaRice. IRRI administers the overall project and is responsible for delivering rice to the Asia region, whereas AfricaRice is responsible for coordinating the Africa side.

12 http://irri.org/partnerships/networks/cure/cure-news/new-drought-tolerant-rice-varieties -released-for-the-western-mid-hills-of-nepal as well as http://irri.org/news-events/media -releases/nepalese-farmers-to-enjoy-bountiful-harvest-from-drought-proof-rice (both accessed May 2012).

13 This is the hunger season, during September and October.

14 http://irri.org/partnerships/networks/cure/cure-news/outlasting-drought-with-ghaiya-1-in -upland-nepal (accessed May 2012).

varieties that have high yields under rainfed and drought conditions but also maintain yields when water becomes available (during favorable years or when irrigated). Although the mechanisms of drought tolerance in wheat are only partially understood, some progress in the development of drought-tolerant varieties has been made through selection under drought stress (Ortiz et al. 2008), but success in conventional breeding strategies has been hampered by the polygenic nature of drought tolerance (Khan et al. 2011).

CIMMYT has developed many resynthesized hexaploid wheat lines, obtained by crossing the diploid wild ancestor *Aegilops tauschii* (goat grass) with tetraploid durum wheat (*Triticum turgidum* var. *durum*). These hexaploid varieties have inherited genetic material from the wheat wild relative *A. tauschii,* which provides characters useful for the development of improved tolerance to drought and heat stress (Reynolds, Dreccer, and Trethowan 2007; Ashraf 2010). In multi-site trials, some of these lines have shown yields that were between 8 and 30 percent higher than those of the best local varieties across various environments in Australia (Ogbonnaya et al. 2007).

Improved Varieties—Heat-Tolerant Characters

Recent studies provide evidence that developing wheat and other crops to adapt to high temperatures should be a top priority for plant physiologists and crop breeders (Ciais et al. 2005; Battisti and Naylor 2009; Lobell, Sibley, and Ortiz-Monasterio 2012). Some strides have been made in understanding the effects of heat on crops and yields. It is now known that the sensitivity of crops to high temperatures varies during the life cycle, with flowering being the most sensitive time in plant growth, as heat can disrupt pollination and therefore yields. Furthermore, evidence indicates that heat can accelerate the rate of plant senescence (Lobell, Sibley, and Ortiz-Monasterio 2012).

The commercial availability of heat-tolerant crops is still distant. For rice, progress in breeding has been encouraged by the availability of the full genome. Research based on marker-assisted selection and genetic modification is targeting both the enhanced fertility of flowers at high temperature and the development of varieties with shorter duration to avoid periods of peak stress (Shah et al. 2011). Similar efforts are ongoing for wheat. Although the genetic basis for heat resistance is still unknown, researchers are studying those physiological traits that seem related to adaptation to warmer temperatures (Cossani and Reynolds 2012). In addition, the great variety of genetic material in germplasm banks (landraces, wild relatives, and the like) and the declining costs of genetic and genomic analyses are fueling optimism about identifying the genetic basis of heat-adaptive traits (Cossani and Reynolds 2012).

Crop Protection

Crop production increases stemming from greater access to resources, increased inputs, or many types of improved management practices generally go hand in hand with increased potential for losses due to pathogens, animal pests, and weeds (collectively referred to as "pests") (Oerke et al. 1994; Oerke 2006). Denser crop canopies, shorter intervals between crops, monoculture, and increased fertilizer use often result in higher pest populations. Efforts to intensify agricultural production are therefore incomplete without addressing the concurrent need to invest in crop protection.

Since the early 1960s, the application of herbicides, fungicides, and insecticides has increased 15- to 20-fold, and sales of these agents have jumped 30-fold, about US$30 billion worldwide (Oerke 2006). Although grain production has also doubled over the past 40–50 years, partially as a consequence of changes in crop protection, the overall proportion of crop losses has actually increased (Oerke et al. 1994; Oerke 2006). Depending on the crop, pests are responsible for 25–50 percent or more of global crop losses (Oerke 2006). Losses are particularly devastating in poorer regions of the world, where climates are relatively wet and warm, crops are grown nearly all year or without rotation, crop varieties or landraces are susceptible, and crop protection is absent or of low efficacy (Oerke et al. 1994). Indeed, severe pest outbreaks can be the main cause of starvation in developing countries, especially in areas dominated by subsistence agriculture (Chakraborty, Tiedemann, and Teng 2000; Strange and Scott 2005).

Crop protection is based on a variety of practices and technologies. Cultural practices (tillage, crop rotation, optimal planting windows, and intercropping), plant genetics (pest-resistant or pest-tolerant crop varieties), biological control (organisms typically benign to crops but that attack, parasitize, or outcompete crop pests), and synthetic pesticides are prime examples. Use of crop varieties that are genetically resistant to major pests can effectively protect against substantial losses. Crop breeders face the challenge of developing new forms of resistance when pathogens and arthropods evolve to overcome crop resistance, just as pathogens and arthropods can evolve resistance to pesticides. However, many success stories and cases of long-lived, durable resistance genes have been noted (Bockus et al. 2001). Additionally, breakthroughs in genetics help speed up breeding programs and allow breeders to incorporate multiple desirable traits into crop varieties (Xu and Crouch 2008; Heffner, Sorrellsa, and Jannink 2009). The use of a combination of minor genes for resistance has proven to be a successful strategy against wheat rusts in many areas of the world (Singh et al. 2011).

If multiple options are available, farmers rarely rely on one technique, product, or practice to protect their crops from pest damage. Rather, to maximize pest control, reduce risks, and extend the shelf-life of chemical products and genetic resources, most experts recommend an integrated pest management approach (Krupinksy et al. 2002). According to World Bank (n.d.),

> IPM [integrated pest management] refers to a mix of farmer-driven, ecologically based pest control practices that seek to reduce reliance on synthetic chemical pesticides. It involves (a) managing pests (keeping them below economically damaging levels) rather than seeking to eradicate them; (b) relying, to the extent possible, on non-chemical measures to keep pest populations low; and (c) selecting and applying pesticides, when they have to be used, in a way that minimizes adverse effects on beneficial organisms, humans, and the environment.

Genetic resistance to insect pests and pathogens in crop varieties (when available) is widely regarded as the first line of defense, often in combination with various cultural practices mentioned above and chemical pesticides as back-up when necessary, recommended by expert forecasting, and affordable to farmers.

Chapter 3

Methodology: Choice of Models, Limits, and Assumptions

R epresenting agricultural technologies and their roles in the global agricultural economy requires a framework incorporating many separate
pieces so they can work together. The modeling framework used in this
book is presented in Figure 3.1, which shows how the different modeling components are linked. This framework relies on the combination of DSSAT
(a process-based crop model) and IMPACT (a global, partial-equilibrium,
agricultural sector model).

FIGURE 3.1   Modeling system for estimation of impacts of agricultural technologies

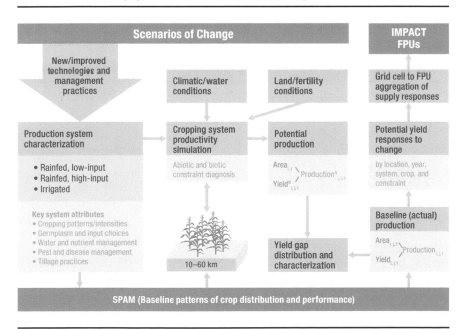

Source: Authors.
Note: FPU = food-producing unit; IMPACT = International Model for Policy Analysis of Agricultural Commodities and Trade;
SPAM = Spatial Production Allocation Model.

Modeling Framework

As a first step, the production systems of maize, rice, and wheat are characterized using a series of global, high-resolution datasets, such as the spatial databases of crop geography and performances, climate scenarios, and soil properties. Based on these gridded data, a baseline of existing dominant management practices and inputs (germplasm, nutrients, supplemental water, and pesticides) are assembled by water management system (e.g., rainfed or irrigated) and by agroecological zone. This baseline is then simulated with high granularity (0.5-degree, or about 60-kilometer, grids) in the process-based DSSAT model separately for rainfed and for irrigated farming systems.

In the second step, alternative agricultural technologies are characterized in DSSAT, again separately for rainfed and for irrigated farming systems. The cropping system productivity simulation assesses whether the specific agricultural technology being evaluated outyields the baseline yield at that specific cell and whether the technology may induce changes in water and nitrogen use, again compared to the baseline. Simulated yields are then aggregated to the level of food-producing units (FPUs). An FPU is the lowest area-input level of IFPRI's IMPACT, which is a global partial-equilibrium agricultural sector model designed to simulate and examine alternative futures for global food supply, demand, trade, prices, and food security.

The crop modeling part of the framework deals in detail with the technology and climate specifications. The DSSAT business-as-usual baseline assumes that the technologies tested in this study are not adopted; instead, the same mix of agricultural practices in use in the baseline period of 2010 is assumed to be maintained across the entire period 2010–2050. The DSSAT baseline simulated yields reflect our best understanding of farmers' management practices, based on a compilation of global datasets, the literature, and our own synthesis of crop model input parameters. More details about the DSSAT baseline are provided later in this section. All technologies assessed in the study, such as ISFM and water harvesting, were implemented in the crop models by adjusting model input parameters or coding the management practice in detail (or both) to reflect how farmers would implement the technology in the field. We simulated the baseline and all technologies under two climate scenarios, the Commonwealth Scientific and Industrial Research Organisation's general circulation model (CSIRO) A1B and the MIROC A1B scenarios.[1] We focused on the MIROC A1B scenario, but differences with the CSIRO A1B scenario are highlighted.[2]

---

1   We call the use of two different global circulation models with one common SRES scenario "scenarios," recognizing that this is not the way that the IPCC uses the term.

2   Details on the two climate scenarios can be found in Appendix A.

The results of DSSAT were then fed into IFPRI's model, IMPACT, using adoption pathways that consider profitability, initial costs and capital, risk-reduction, and complexity of the technology. We then simulated global food supply and demand, food trade, and international food prices, as well as the resulting population at risk of food insecurity, which leads to comparisons of the benefits from different technologies.

DSSAT simulates the main processes of crop growth and can measure changes in yields as affected by changes in geographic location, varietal use (that is, crop varieties), soil, climate, and management (that is, agricultural technologies and practices). The advantages and disadvantages of process-based models (versus, for example, statistical models) have been described in several papers (Lobell and Burke 2009, 2010; Lobell et al. 2011). These simulation models often provide only point estimates rather than intervals, are often calibrated to temperate regions (or wherever trial data are available), and do not include all potentially relevant biological processes. Also, they are dependent on inputs describing cultivar characteristics, management practices, soil properties, and initial conditions for all these parameters (Lobell and Burke 2010). On the plus side, process-based models rely on decades of research on crop physiology, agronomy, soil science, and other disciplines, and the parameters and data used to model the processes (that is, the interactions among crop, soil, and climate) have relatively well-understood physical meanings.

In contrast to regression-based approaches or statistical models, process-based models like DSSAT allow for an explicit representation of the constituent processes (soil, water, plant, and so forth) and explicitly facilitate the simulation of new cultivars, management practices, or both. For example, a process-based model of maize will have a specific parameter controlling the sensitivity of crop maturation due to photoperiod, and this will change for every maize variety and at different latitudes. Adjustment of this parameter can provide insight into the changes in yields or other parameters stemming from adoption of alternative varieties. Additionally, there is the practical aspect of obtaining sufficient geographic coverage that can match the scope of a global economic model. Generating global estimates with regression models would require adequate yield data from all geographic regions covered by the study. Process-based models allow us to sidestep this requirement by accepting the universal way the functions of plants and crops are represented and calibrated.

All the strengths and weaknesses of process models notwithstanding, the defining characteristic of the approach used by our team at IFPRI is the link between DSSAT and IMPACT. Combining the two models allows simulation of how disaggregated changes in agricultural productivity may affect global and regional production, global food prices, and trade flows, as well as calorie availability and malnutrition levels in developing countries and across the world.

An overview of the DSSAT and IMPACT models and details on how the crop modeling results are incorporated into IMPACT can be found in Appendix A. In the rest of this chapter, we illustrate the general approach of the study and the assumptions behind the technology scenarios.

DSSAT and IMPACT Baselines

To estimate the impact of the alternative agricultural technologies in DSSAT and subsequently in IMPACT, a plausible counterfactual baseline (or business-as-usual management practice) scenario must be defined.

DSSAT Baseline Management Scenario (Business-as-Usual)

The baseline management scenario is a reflection of current implementation of technologies assessed and assumes that farmers are consistently not adopting any of the specific technologies assessed in this study throughout the simulated period of 2010–2050. For each agroecological zone, a representative cultivar with high-yielding characteristics was chosen for each crop, assuming the adoption of improved varieties will precede the adoption of technologies. The geographic distribution of representative cultivars in the crop modeling analysis was assumed static over the simulation period (that is, the extent of improved varieties does not change over time). This was to keep the impact of technologies—whose assessment is the primary topic of this analysis—separate from the possible interactive effects among cultivars, technologies, and environment that can make the impact of technology itself difficult to analyze. For both rainfed and irrigated systems, suboptimal planting density was assumed, with details based on literature reviews, consultation with experts, and sensitivity analyses in DSSAT, using a wide range of plant densities. In addition, we assumed suboptimal planting windows in both rainfed and irrigated systems; these are defined as a fixed, narrow window around the middle of the most likely planting month (Nelson et al. 2010). The baseline includes an inorganic nitrogen fertilizer application rate that is specific to each region, input system, and crop and was obtained by calibration of simulated raw yields to FAOSTAT yields. A simple, calendar day–based split application of inorganic fertilizer scheme was used (that is, 50 percent on planting at 5-centimeter depth, and the other 50 percent banded on the surface 20 days after planting) for maize and rice; for wheat, all fertilizer was applied on the day of planting. Where irrigation is adopted, furrow irrigation was used as the baseline irrigation technology. For tillage practice, conventional tillage (that is, plowed 10 days before planting with a moldboard plow at 20-centimeter depth causing 100 percent soil disturbance) was used for maize and wheat; for paddy rice, puddling was used in the baseline as well as across technologies. Most (90 percent) of aboveground crop residues were removed from the field right after harvest. The model was set up to simulate seasonal crop growth and harvest sequentially over the simulation period of 43 years. The initial soil moisture level before planting was set to reflect 50 percent and 100 percent of field capacity for rainfed and irrigated systems, respectively, for the first year; in subsequent years, the model carried over the soil water and nitrogen balances from the previous season. The first 3 years of results were discarded as a spin-off period to balance the system-specific soil water and nitrogen status at the beginning of cropping season.

The baseline scenario in the crop modeling analysis assumes no changes in technology adoption between 2010 and 2050. However, the simulated yields change over time because of the estimated changes in soil fertility and their interactions with projected changes in future climate conditions.

IMPACT Baseline

IMPACT uses a system of linear and nonlinear equations to approximate the underlying production and demand relationships of world agriculture. The world's food production is disaggregated into 115 countries and regional groupings, and 126 hydrological basins. Overlay between these two groups creates 281 FPUs, which are the basic units of analysis (Rosegrant and IMPACT Development Team 2012). The IMPACT baseline includes gradual improvements in crop yields over 2005–2050 in response to continued changes in supply and demand, such as those caused by population and economic growth, the resulting changes in diets, and global food trade. The main supply-side drivers to the IMPACT baseline are elasticities, yield and technology growth assumptions, and area growth assumptions, which are exogenous to the model. Data for these components come from historical data trends adjusted by expert opinion. The IMPACT technical document provides the source references for the initial historical data, including the data for population and gross domestic product, which are the demand side drivers (Rosegrant and IMPACT Development Team 2012). Food supply, demand, and trade data for the baseline period (2005, an average of 2004–2006) are derived from the FAOSTAT database. Baseline commodity price data are chiefly derived from the World Bank pink sheets (for more details, see Appendix 8 in Rosegrant and IMPACT Development Team 2012). Population statistics are drawn from the UN Population Division's *World Population Prospects* (UN 2011). Regional income growth is based on a World Bank study (Margulis 2010) and has been updated for SSA and South Asian countries. The effects of climate change on crop production enter the model through changes in area and yield, again estimated through DSSAT as well as an associated Global Hydrology Model to account for the effects of climate change on irrigation water availability. This methodology allows for baseline scenarios specific to different climate change futures.

In IMPACT, the availability of kilocalories per capita is derived from the amount of calories obtained from commodities included in the IMPACT- Food model as well as calories from commodities outside the model (Rosegrant and IMPACT Development Team 2012). The share of malnourished children under the age of five is estimated based on average calorie availability per capita and day, female access to secondary education, the ratio of female to male life expectancy at birth, and the share of people with access to clean water (Rosegrant and IMPACT Development Team 2012). Observed relationships among these factors are used to build a semi-log functional mathematical model to estimate the number of malnourished children. The estimated equation is based on a cross-country regression developed by Smith and Haddad (2000).

The share of people at risk of hunger (that is, the share of the total population at risk of food insecurity) is calculated based on an empirical correlation between the share of malnourished people within the population and the relative availability of food (Rosegrant and IMPACT Development Team 2012). The calculation is adapted from Fischer et al. (2005).

Additional technical details on the IMPACT baseline can be found in Appendix A. Detailed specifics on how the IMPACT baseline is constructed and how calorie availability per capita per day, the number and share of malnourished children, and the population at risk of hunger are calculated can be found in the IMPACT technical document.[3] Additional information on the use of baselines can be found in Nelson et al. (2010).

Agricultural Technologies Assessed

Following extensive consultations, 11 technologies were chosen for detailed assessment. The technologies cover a broad range of traditional, conventional,

---

3 Available for download from http://www.ifpri.org/book-751/ourwork/program/impact-model.

and advanced practices with some proven potential for yield improvement and potential for wide geographic application. Several of these technologies have already been partially adopted in some parts of the world, such as no-till or drip irrigation. Others are in the final stages of development and field trials. All of them can be rolled out in one form or another across large agricultural areas if appropriate investments, support policies, and institutions are put in place. The costs of these measures (which could differ across regions) have not been explicitly accounted for in this study due to data limitations. The technologies are

1. no-till (minimum or no soil disturbance, often in combination with residue retention, crop rotation, and use of cover crops);

2. integrated soil fertility management (combination of chemical fertilizers, crop residues, and manure/compost);

3. precision agriculture (GPS-assisted delivery of agricultural inputs and low-tech management practices that aim to control all field parameters, from input delivery to plant spacing to water level);

4. organic agriculture (cultivation with exclusion or strict limits on use of manufactured fertilizers, pesticides, growth regulators, and genetically modified organisms);

5. water harvesting (water channeled to crop fields from macro- or micro-catchment systems, or through the use of earth dams, ridges, or graded contours);

6. drip irrigation (water distributed by a small discharge directly around each plant or to the root zone, often using microtubing);

7. sprinkler irrigation (water distributed under pressure through a pipe network and delivered to the crop by overhead spraying through sprinkler nozzles);

8. heat tolerance (improved varieties showing characters that allow the plant to maintain yields at higher temperatures);

9. drought tolerance (improved varieties showing characters that allow the plant to have better yields compared to regular varieties due to enhanced soil moisture uptake capabilities and reduced vulnerability to water deficiency);

10. improved nitrogen use efficiency (varieties showing enhanced NUE); and

TABLE 3.1   Summary of technologies simulated in DSSAT and IMPACT

Technology	Abbreviation	Crop	Rainfed/irrigated
No-till	NT	Maize, wheat	Rainfed and irrigated
Integrated soil fertility management	ISFM	Maize, rice, wheat	Rainfed and irrigated
Precision agriculture	PA	Maize, rice, wheat	Rainfed and irrigated
Organic agriculture	OA	Maize, rice, wheat	Rainfed and irrigated
Water harvesting	WH	Maize, wheat	Rainfed
Drip irrigation	DRIP	Maize, wheat	Irrigated
Sprinkler irrigation	SPRK	Maize, wheat	Irrigated
Heat tolerance	HT	Maize, rice, wheat	Rainfed and irrigated
Drought tolerance	DT	Maize, rice, wheat	Rainfed
Nitrogen-use efficiency	NUE	Maize, rice, wheat	Rainfed and irrigated
Water harvesting + integrated soil fertility management	WH +FM	Maize, wheat	Rainfed
No-till + water harvesting	NT+WH	Maize, wheat	Rainfed
No-till + precision agriculture	NT+PA	Maize, wheat	Rainfed and irrigated
No-till + heat tolerance	NT+HT	Maize, wheat	Rainfed and irrigated
No-till + drought tolerance	NT+DT	Maize, wheat	Rainfed
Drought tolerance + heat tolerance	DT+HT	Maize, wheat	Rainfed
Crop protection—diseases	CP-D	Maize, rice, wheat	Rainfed and irrigated
Crop protection—insects	CP-I	Maize, rice, wheat	Rainfed and irrigated
Crop protection—weeds	CP-W	Maize, rice, wheat	Rainfed and irrigated

Source: Authors.

Note: DSSAT = Decision Support System for Agrotechnology Transfer; IMPACT = International Model for Policy Analysis of Agricultural Commodities and Trade.

11. crop protection (effects of chemical treatment against diseases, insects, and weeds).[4]

Table 3.1 provides a summary of these technologies together with information on which crops they were applied to, type of farming systems, and type of watering system (irrigated or rainfed).

Only a few technologies in this list could technically not be adopted simultaneously, for example, drip irrigation and sprinkler irrigation, but both can still be used in separate sections of the farm or in different cropping

---

4 The use of breeding or other techniques to develop crop resistance to diseases and insects is a desirable management strategy, but global data were not available for its assessment. Effective resistance would produce gains in yield comparable to effective pesticide use.

seasons. Although we could not assess all possible combinations as part of this study, we tested six sample crop technology combinations. Several of these combine enhanced crop management practices with advanced breeding strategies:

1. integrated soil fertility management + water harvesting,

2. no-till + water harvesting,

3. no-till + precision agriculture,

4. no-till + heat tolerance,

5. no-till + drought tolerance, and

6. drought tolerance + heat tolerance.

Moreover, we explored one multiple-adoption scenario of all the technologies using IMPACT alone. This simulation of stacked technologies also made it possible to show the marginal contribution of each technology to the overall impacts of the stacked technologies.

Technology Implementation in DSSAT

Each of the promising technologies and combinations thereof required individual specification of different details in the crop modeling simulations and used high-yielding varieties. The description of a technology identifies the aspects that specifically distinguish that technology from the DSSAT baseline. Therefore, if the description of a technology (for example, no-till, discussed below) provides no specifics on nitrogen inputs, it is implied that the nitrogen application rate is the same as in the baseline.

Furthermore, we assume that all technologies, with the exception of no-till, are not yet adopted in the baseline scenario. As a result, we may overestimate the yield results for the technologies in some regions where they are already in use to some degree.

No-till

Simulation of no-till was set up as the opposite of the baseline tillage practice: the tillage option was switched off. To minimize soil disturbance, a seed planting stick was simulated as the planting method. For fertilizer application, a deep injection method was simulated under no-till. No-till was applied globally, except for six countries (Argentina, Australia, Brazil, New Zealand,

Paraguay, and Uruguay),[5] where the no-till practice is already widely applied on the majority of cropland. Potential for expansion of no-till in the six countries was considered negligible. We do not apply no-till to rice. Given that there is substantial no-till in North America and a few other countries, we likely overestimate returns from this practice in these countries.

Integrated Soil Fertility Management (ISFM)

ISFM was applied globally and was implemented by applying organic amendment in addition to the inorganic fertilizer applications defined in the baseline management scenario. The site-specific organic manure rate was based on Potter et al. (2010; data downloaded from http://www.earthstat.org).[6] We assumed the organic manure came from livestock and contained 1.4 percent nitrogen. The total amount was applied monthly during the fallow period (after harvesting–before planting) at the rate of 1 metric ton/hectare. The average rate of manure application, in terms of its nitrogen content, was between 9 kilograms [N]/hectare in Southern Africa and 46 kilograms [N]/hectare in South Asia. The rate of inorganic fertilizer application is the same as the reference case (business-as-usual); however, the application scheduling is optimized based on the growth stage of each crop to minimize nitrogen stress during flowering and grain filling.

Precision Agriculture (PA)

PA is applied globally for both irrigated and rainfed systems. The multiple effects of PA are implemented using three components: (1) higher/optimum planting density for each crop, predetermined based on the results of sensitivity analyses; (2) enhanced inorganic fertilizer application scheduling based on the growth stage of the crop (same as ISFM); and (3) optimum planting window, which assumes a plant-available soil water content of 100 percent field capacity, assuming a 25-millimeter rainfall event on the planting date.

---

5 Countries in North America, particularly the United States, were not included in the list of countries where no-till is largely adopted, as their current adoption level of no-till is not as universal as in the other six countries and because adoption continues to increase in North America (Horowitz, Ebel, and Ueda 2010).

6 The assumption of livestock as the source of organic manure introduces a source of inaccuracy in the IMPACT simulations (see Chapter 5) that assume a uniform diffusion path across regions. Many smallholders in some regions do not have sufficient livestock to produce the manure required for adoption of ISFM, many use animal dung as fuel rather than as fertilizer, and green manure availability is limited as well. Therefore, the applicability of this constraint varies across regions. The cost of transporting and applying manure to places of ISFM expansion was a major reason for the low adoption ceiling of 40 percent (see Table 3.3).

Organic Agriculture (OA)

OA requires that neither chemical protection agents nor inorganic fertilizers be applied and that genetically modified varieties are not used. DSSAT has no explicit mechanism to simulate biotic constraints (for example, pests or plant diseases). We therefore approximate OA systems by using an estimated site-specific factor of yield reduction to discount the baseline yields simulated by DSSAT. We retrieved the site-specific factors of organic-to-conventional crop yield ratios (OCRs) from the Seufert, Ramankutty, and Foley (2012) meta-analysis, which compares 315 observations of yields from OA and conventional agriculture from 62 sites globally, published during 1980–2010. The meta-analysis considers multiple dimensions of field management, such as nitrogen input, best management practice, soil pH, irrigation, the time since conversion to OA, and country development. The meta-analysis provides detailed contextual comparisons of OCRs for different crops.

We separate OCRs of crops, irrigation, development status, and continent[7] and apply them to our estimates of harvested areas to approximate the yield ratio for each of 72 classes: 6 continents × 2 watering choices (irrigated/rainfed) × 2 country traits (developed/developing) × 3 crops (maize/wheat/rice). We use these data to scale the baseline yields to estimate yields of OA (Figure 3.2 and Appendix B).

De Ponti et al. (2012) analyze global yield gaps in OA and conventional agriculture across crop groups and regions based on 362 data records, concluding that the OCR is about 80 percent globally. The study provides regional breakdowns of OCRs, but the data are aggregated regionally across multiple crops and are not useful for our purposes. Moreover, they did not provide details of field management, so that it was impossible to understand how OCRs were influenced by such input differences as rainfed versus irrigated systems.

Water Harvesting

Water harvesting is applied globally in rainfed agriculture areas. A two-stage simulation approach is implemented. The simulation is first run without water harvesting. From the simulation output, the phenology of each season (planting, flowering, and maturity dates) as well as runoff from the field are recorded. Assuming some in situ water storage potential that captures runoff, the simulation output is further analyzed to determine when supplementary irrigation is most needed (for example, soon after germination and before flowering, when accumulated runoff was greater than 25 millimeters), and how much of the

---

7 Other aspects did not have sufficient geographic coverage.

FIGURE 3.2 Aggregated average organic-to-conventional crop yield ratios (OCRs)

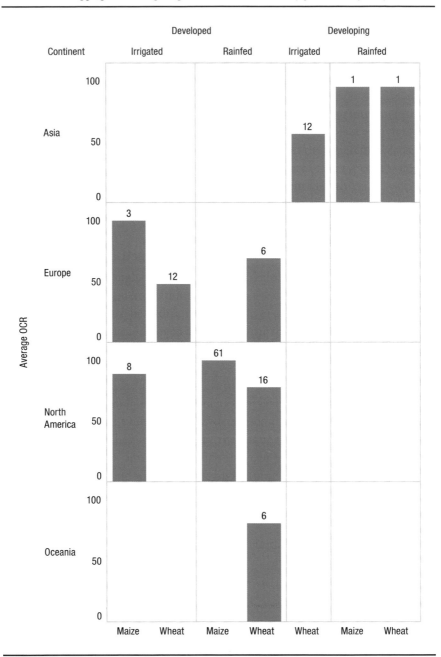

Source: Seufert, Ramankutty, and Foley (2012).

Note: Empty spaces indicate there were no data available. Number above each bar indicates the number of records. Seufert, Ramankutty, and Foley (2012) did not include rice.

harvested water would be available from the in situ storage device (for example, 80 percent of runoff was available to the field as supplementary irrigation). The simulation is then run again with the supplementary irrigation applied.

Advanced Irrigation Technologies: Drip Irrigation and Sprinkler Irrigation

Two advanced irrigation technologies are applied on areas that are already irrigated: drip irrigation and sprinkler irrigation. Furrow irrigation is assumed as the baseline management practice. Improved irrigation technologies were simulated with irrigation efficiency coefficients based on Howell (2003): furrow irrigation as the baseline method with an efficiency of 0.65, sprinkler irrigation with an efficiency of 0.75, and drip irrigation for maize and wheat with an efficiency of 0.90. We do not simulate drip irrigation for rice. To implement the irrigation technologies, DSSAT's automatic irrigation method was first used to identify when and how much water is needed during the cropping season. Irrigation is considered necessary when precipitation is insufficient to meet the predefined requirement of 40 percent of field capacity to a soil depth of 30 centimeters. The simulation was then run again with manual irrigation, following the same scheduling as the automatic irrigation, but with adjustments for irrigation efficiency depending on the irrigation technology and adjusted for the regional river-basin-level parameter called "irrigation water supply reliability." This parameter is based on Rosegrant, Cai, and Cline (2002) and reflects relative irrigation water scarcity at the river basin level.

Heat Tolerance

Heat tolerance characteristics are applied globally on both irrigated and rainfed systems. The warming temperature trend under climate change scenarios in general shortens crop durations such that cereal crops have less time to fill the grain and consequently decreases yield. We hypothesize that future heat-tolerant varieties are being bred to maintain the same crop duration and phenology under warming climate conditions. Ideally, if such varieties already exist, detailed cultivar growth and phenology characterization of the heat-tolerant variety can be conducted and implemented in the crop modeling framework. Because such detailed information is lacking, we instead simulate the effects of the heat-tolerant traits by manipulating the generation of future daily weather data: the parameters are set such that the stochastically generated daily weather data maintains the same mean of daily minimum and maximum temperatures as the current climate (circa 2000). Other weather variables, such as solar radiation and rainfall, reflect the climate change scenarios.

Drought Tolerance

Drought tolerance characteristics were applied in rainfed areas for all crops. To simulate drought-tolerant traits, we focused on the enhancement of roots. The improved root volume was simulated by increasing the soil root growth factor parameter[8] of each layer. Enhanced water extraction capability is also simulated by lowering the lower limit parameter[9] in the soil profile. For maize, the reduced sensitivity of hybrid varieties to the anthesis-to-silking interval (ASI) was also simulated.[10] To implement the ASI mechanism and its sensitivity, the existing CERES-Maize[11] model was modified, and the differential sensitivity to ASI was developed as a cultivar trait. There are other types of potential mechanisms and breeding strategies not implemented in this study; we assumed their benefits to crops would be reasonably similar to those resulting from having more access to water, as implemented in this study.

Improved Nitrogen-Use Efficiency

Improved NUE through breeding was applied globally on both irrigated and rainfed systems. Parameterization of NUE in the crop models includes manipulation of each crop to be additionally capable of producing more biomass per the same amount of nitrogen available in the soils. Specifically, this was differently implemented in each crop model because of their differences in the species and cultivar characteristics. We assume that the crops with NUE would hypothetically be less susceptible to the nitrogen stress (the parameter used for maize, which indicates the fraction of leaf area deteriorated due to age under 100 percent nitrogen stress, decreased from 0.050 to 0.045), would recover faster from the nitrogen stress (the parameter used for wheat, which indicates the fraction of nitrogen deficit that can be filled per day, increased from 0.05 to 0.10), and would possess higher potential of grain weight under ideal growing condition (the parameter used for rice, which indicates the

---

8   The values of this parameter indicate the root growth in each soil layer across the soil depth. A value of 1 indicates root growth in a layer has no limitation. A value of 0 indicates root growth is constrained.

9   The parameter values indicate the amount of volumetric moisture content in the soil layer after roots have extracted all the water they are able to extract. It is measured in cubic millimeters of water per cubic millimeter of soil.

10 ASI refers to the time between the emergence of the male and female flowers on the maize plant. Synchrony between emergence of the flowers is disrupted by drought, and therefore fertilization is reduced. In a drought-resistant plant the synchrony of emergence is re-established.

11 CERES maize is the actual name of the maize model included in DSSAT. See Jones and Dyke (1986).

single grain weight under stress-free environment, increased by 15 percent for each variety). These parameters can be found in the species and cultivar files of DSSAT (MZCER045.SPE for maize, WHCER045.SPE for wheat, and RICER045.CUL for rice).

Crop Protection

For the purpose of this book, we define crop protection as any approach that effectively reduces damages caused by pests and therefore benefits crop yields. However, to analyze the effects of crop protection on yields, the data in this study were derived specifically from reported chemical control of rice, wheat, and maize pests, because these data are currently more accessible and harmonized across different regions of the world (Oerke et al. 1994; Oerke and Dehne 2004; Oerke 2006). It should be noted that these data do not necessarily reflect the wider context of emerging pests and diseases, pest adaptation, ecological interactions, and unintentional consequences as a result of pesticide applications (for example, brown plant hopper in Asia is a secondary problem due to insecticide spraying for leaf-feeding insects) (Way and Heong 1994; Bottrell and Schoenly 2012). Some potential concerns related to pesticide use were beyond the scope of this project, although they are important considerations in more complete evaluations of the overall costs and benefits of pesticide use. We did not consider the misuse of pesticides; the consequences of inappropriate pesticide policies or regulations; and the effects of pesticides on ecosystem health, farm workers, downstream settlements, and consumer health.

Despite the limitations in data availability, we have set out to estimate the possible effects of crop protection on yields at the regional and global levels. We focus on the effects of control of insects, pathogens, and weeds. Given the large uncertainty and the current knowledge gaps on pest distribution under climate change, their rate of adaptation to new conditions, the effects of the climate on the plants' natural defenses, and the interaction between pests and protection measures, we provide scenarios as a starting basis for policy discussion. Although data are derived specifically from reported chemical control of rice, wheat, and maize pests, the reader should note that our use of chemical pesticide data is meant to illuminate the importance of crop protection on yields as part of any good management system and technology adoption in general. It is not our intent to evaluate or recommend a policy of worldwide adoption of pesticides. Furthermore, effective crop resistance to pests would produce gains in yield comparable to effective pesticide use.

TABLE 3.2   Targeted PAWs for wheat, maize, and rice

	Wheat		Maize		Rice	
Item	Common name	Scientific name	Common name	Scientific name	Common name	Scientific name
Pathogen	Stripe (yellow) rust	*Puccinia striiformis*	Grey leaf spot/ common rust	*Cercospora zeae-maydis/ Puccinia sorghi*	Rice blast	*Magnaporthe oryzae/ M. grisea*
Arthropod	Russian wheat aphid	*Diuraphis noxia*	European corn borer	*Ostrinia nubilalis*	Yellow and striped stem borers	*Scirpophaga incertulas/ Chilo suppressalis*
Weed	Wild oat	*Avena fatua*	Johnson grass/ lambsquarters	*Sorghum halepense/ Chenopodium album*	Barnyard grass	*Echinochloa crus-galli*

Source: Authors.
Note: PAW = pathogen, arthropod, weed.

To fit in the scope of this book, our pest prevalence modeling is limited to approximately three species each for wheat, maize, and rice, including one fungal pathogen, one arthropod (insect) pest, and one weed (collectively referred to as PAW) (Table 3.2). Our general assumption is that each species chosen for our analysis represents a key species from the pool of pests that are currently treatable with pesticides and are important to the crop in terms of critical yield losses and global significance. Clearly, other pests, pest complexes, or injury profiles[12] are often equally or more important in certain regions of the world. However, the objective of this book is to carry out a preliminary analysis linking climate change and farming technologies with key pests and crop protection.

Integrating potential PAW prevalence with crop protection technologies (pesticides, fungicides, and herbicides) into the crop modeling framework required the following steps. First, we estimated the regionally aggregated actual (with or without crop protection) and potential (with crop protection technologies fully adopted and effectively used) yield losses from the three different types of biotic constraints (for details, see Appendix A). Second,

---

12 Injury profiles are defined as the combination of injury levels caused by the multiple pests (pathogens, insects, and weeds) that affect a crop during a growing cycle (Savary et al. 2000, 2006).

the regionally aggregated actual and potential yield losses were spatially dis-aggregated using global prevalence surface (map) data for each pest, disease, and weed (presented in Appendix C) as a primer. Third, at the grid cell–level, the potential yield impact (that is, the difference between the potential and actual yield losses) was computed and used to increase the baseline yield. This method implicitly assumes that the business-as-usual baseline yield is already calibrated with the actual level of yield that takes into account the actual yield loss from the biotic constraints. As a result, the application of crop protec-tion technology can increase the yield by effectively removing the biotic con-straints. In the above context, "actual yield" refers to an estimate of the yield that would be achieved in the field as a result of management choices, rather than to observed yield.

Although we only evaluated representative PAWs rather than all impor-tant biotic stressors, the yield loss stemming from all members of a PAW group was used to estimate yield loss for the representative species. For exam-ple, the representative arthropod pest for maize was treated as producing maize yield loss equivalent to available estimates of maize yield loss from all arthropod pests. Similarly, although a representative PAW species might not be present in all parts of the world (because of limitations on dispersal or effective quarantine), we evaluated global distributions based on climate favorability. Because the maps of PAW prevalence are based solely on climate favorability, it is important to keep in mind that environmental conditions that favor a pest can also favor the natural enemies of a pest. This is appar-ently the case for rice insect pests in some locations, as reflected in the IRRI recommendation against insecticide use (Way and Heong 1994; Bottrell and Schoenly 2012). These two methodological steps were intended to provide a better approximation to yield loss for the whole group of PAWs for a particu-lar crop.

Calculation of Yield Impact in DSSAT

As our primary concern is with the effect of promising technologies, our discus-sion of the results focuses on yield changes between a baseline case (business-as-usual) without technological improvements and the outcomes of technology adoption in the end period (2041–2050). Yields were simulated on a yearly basis between 2010 and 2050, and the start and end periods were used to com-pute the yield impact. To avoid very large and very low yields in individual pixels that could skew final results, we apply a yield minimum of 500 kilogram/hectare as the lower boundary and a maximum yield gain of 200 percent as the upper boundary for all crops in DSSAT (these boundaries were rarely binding).

The measured yield impact from adoption of a technology under a particular climate scenario and for a given grid cell is summarized as follows:

$$\text{Yield Impact (\%)} = \max\left(\frac{\textit{Yield}_{\text{TECH}} - \textit{Yield}_{\text{BASU}}}{\textit{Yield}_{\text{BASU}}} \times 100\right)$$

where $\textit{Yield}_{\text{TECH}}$ is the model-estimated yield with a specific technology, and $\textit{Yield}_{\text{BASU}}$ is the model-estimated business-as-usual yield with the baseline management scenario (that is, without adopting the specific technology).

Note that the yield impact becomes effectively zero when a yield with technology results in a lower yield than the one without technology adoption. For the purpose of this book, we assume that only positive yields will lead to potential adoption of technologies: pixels with negative yields from alternative technologies are assumed to continue with baseline crop management practices.

Specification of Adoption Pathways of Selected Agricultural Technologies in IMPACT

Technology adoption is an integral phase in the process of agricultural R&D. Any technology is only as good as how well it is adopted and implemented on the ground—in farmers' fields. Even at the conceptual stage of technology development, its adoption pathway is an essential consideration. Technology adoption has two dimensions—technical and socioeconomic feasibility. Technical feasibility relates to the suitability of the technology to the biophysical and agroclimatic conditions of the targeted farms. In contrast, socioeconomic feasibility relates to the appropriateness of the technology with respect to the economic (for example, costs and profitability, input and output prices) and social (for example, culture; taste and preferences; farmers' skills, attitudes, and attributes) environments.

Technical Feasibility

The DSSAT crop model represents the technical feasibility dimension of adoption. By simulating a baseline and alternative technologies, it identifies suitable cropland where yield impacts for these technologies are positive compared to the baseline technology. We assume full (100 percent) adoption of the technologies on technically feasible farms, that is, on farms where yields under the technologies assessed are higher.

Socioeconomic Feasibility

Most of the concerns, studies, and measurements of technology adoption relate to the socioeconomic feasibility of adoption rather than technical feasibility. This is because in reality, in many technically suitable areas, there is no guarantee that a technology would be adopted on farmers' fields. The adoption profiles developed here attempt to reflect, to some extent, the socioeconomic feasibility of the selected agricultural technologies.

The literature on technology/innovation adoption pathways categorizes the major determinants of adoption as

1. farmers/farm characteristics;

2. technology characteristics;

3. the intensity of dissemination and extension activities; and

4. the institutional, policy, and infrastructure situation in the area.

For technologies that have been disseminated for some time and are already adopted on the ground (ex post) technology adoption studies usually focus on determinants 1 (by comparing farmer adopters and nonadopters and the characteristics of their farms), 3 (by examining visits of extension workers, farmers' training activities, extension expenditures, and so forth), and 4 (by studying the availability of roads, credit services, farmers' organizations, and the like). However, this methodology cannot be directly used for ex ante estimations of technology adoption for those technologies that are yet to be developed or disseminated. For this kind of adoption study, the relevant determinants are the characteristics of the technology. This book thus focuses on determinants 2 and 3. The intensity of dissemination and extension activities is linked with adoption, as farmers who are exposed to the technology—whether through extension activities or self-choice—are more likely to adopt the technology (see Diagne 2006).

Shape, Function, and Parameters of Agricultural Technology Adoption

Fitted technology adoption functions in ex post studies commonly follow an S-shaped curve (s-curve). Cumulative normal and logistic functions are the most commonly used algebraic form (Griliches 1957). It is also intuitive for adoption to proceed slowly at first, followed by acceleration with quick spreading of the technology to potential adopters and then by slow deceleration to a saturation point.

The rate of technology adoption has two components: speed and ceiling. The ceiling is the maximum level (or percentage) of farmers adopting the technology. The speed is the time it takes for the technology to be fully adopted by those who chose to adopt it, that is, the time to reach the ceiling — before plateauing (that is, approaching an asymptote) or declining (that is, disadoption due to obsolescence).

Specification of Adoption Pathways

For this book, four members of our research team prespecified the numerical values of the adoption pathway for each of the selected agricultural technologies. These team members relied on their own expertise; evidence gathered throughout this 3-year study and from related studies; and the results of an online survey collecting input from 419 experts on maize, rice, and wheat technologies. The four experts filled in a questionnaire with questions focusing on four major technology characteristics: profitability, initial costs, risk-reduction ability, and complexity—relative to the baseline technology. The first three characteristics reflect the project investment parameters (profitability, investments, risk-reduction ability) commonly used in business planning. The final characteristic, complexity, defines the likelihood that farmers are able to apply the technology correctly (Batz, Janssen, and Peters 2003). These criteria guided the researchers in arriving at a consensus on the likely values for global upper bounds of adoption of each technology and its speed of diffusion, which in turn were used to estimate the respective logistic functions.

Table 3.3 lists the ceiling values of the adoption pathways for the selected agricultural technologies, which were used in the IMPACT simulations. In addition, we specified that all technologies would reach the maximum adoption levels over 30 years. The mixed-technology ceilings were estimated as the intersection (product) of their corresponding single technologies.

To simplify calculations and comparative analyses, we used several assumptions and simplifications during the specification of the adoption pathways. Specifically, we assume that regional and crop variations are influenced by biophysical conditions that are already reflected in DSSAT results and that although socioeconomics conditions are site-specific, their influence is captured through changes in the ceiling value of the technology. Also, the four researchers who participated in the adoption profile development by independently filling out a survey placed the speed of adoption at 25–35 years. We averaged this

TABLE 3.3   Ceilings of technology adoption pathways (%)

Technology	Ceiling
Single technologies	
Drought tolerance	80
Heat tolerance	75
Nitrogen-use efficiency	75
No-till	70
Integrated soil fertility management	40
Water harvesting	40
Drip irrigation	40
Sprinkler irrigation	40
Precision agriculture	60
Crop protection—diseases	50
Crop protection—weeds	50
Crop protection—insects	50
Combined technologies	
Drought tolerance + heat tolerance	60
No-till + drought tolerance	56
No-till + heat tolerance	53
No-till + precision agriculture	42
No-till + water harvesting	28
Water harvesting + integrated soil fertility management	10

Source: Authors.

to 30 years, which is in the mainstream of estimates of full diffusion in ex post studies of agricultural technologies and agricultural research investments (see, for example, Diagne 2006; Alene et al. 2009; Pardey and Pingali 2010).

We consider the implementation of adoption profiles to be a second-best alternative to address the important topic of technology cost. We do believe that more research has to be done in this area.

Sensitivity Analysis of the Adoption Profile

To test the robustness and stability of the adoption profile chosen—especially in the context of the strong assumptions imposed on the specification—we implemented a pessimistic scenario, which was parameterized as 80 percent of the adoption ceiling, while adoption speed was held constant. We also implemented a more optimistic adoption scenario using 120 percent of the adoption ceiling values used in this book. We find that the chosen adoption profile is robust in estimating the transmission of yield impacts from DSSAT to IMPACT—the rankings of the technologies are maintained in each adoption profile scenario (see Table A.4 in Appendix A). The adoption profile is also stable in the range of 80–120 percent for all technologies for the three crops—lower adoption ceilings (80 percent of the rate used in this book) have lower yield effects, whereas higher adoption ceilings (120 percent of the ceiling values used in the book) have higher yield effects, with the yield effects of the chosen adoption profile consistently between the two limits. For example, for heat-tolerant varieties of maize, the agricultural technology with largest yield impact for this crop, the 80, 100, and 120 percent of ceiling values shown in Table 3.3 lead to yield impacts of 13, 16, and 19 percent, respectively. For crop protection (insects) for maize, the respective yield impacts are 2.2, 2.6, and 3.1 percent. Thus, the yield outcomes change in the expected direction.

These results indicate that the adoption profile specification is robust and stable to a wide range of changes in the parameters specified. The sensitivity analysis shows that the comparative analyses of the different technologies with respect to the order and direction of their yield effects (and their consequent effects on prices and food security) are stable across a range of adoption profiles.

Assumptions and Limitations

Model simulations are not predictions of the future; rather, they represent estimates of possible futures given assumptions about climate, baseline growth, and other factors contained in the scenarios that have been designed. Because this study tests the potential of specific technologies under climate change, our scenarios are characterized primarily by assumptions about the type and characteristics of the technology adopted (in DSSAT and IMPACT), by the rate of the adoption of these technologies in different regions of the world (in IMPACT), and by assumptions about the future climate. Moreover, the results of the simulations are also affected by assumptions and underlying data embedded in the baseline scenarios of both models in use (DSSAT and IMPACT).

Additional assumptions are listed in this section. By running technology scenarios on improved varieties, such as those with drought and heat tolerance and improved NUE, we implicitly assume that these improved traits will be available for adoption during 2010–2050. Similarly, we assume adoption of sprinkler irrigation and drip irrigation for cereals, even though they are currently not commonly adopted. In DSSAT, we assume no changes in the extent of global cropland (or share of rainfed and irrigated areas) between 2010 and 2050. Because DSSAT is a purely biophysical model, its power resides in estimating changes in production outcomes and not in determining direct land-use changes. Farmers consider profitability rather than yield, but this could not be simulated in DSSAT. We attempt to reflect profitability differences to some extent in the adoption pathways, but these are at a global level rather than a pixel level and thus negatively affect accuracy.

In contrast, by using the yield effects simulated through DSSAT as inputs, IMPACT can assess the combined effect of productivity changes and changes in population, land use, and food trade, and it can produce results on broad changes in cropland under the three crops used in this study (as well as for all other key food crops).

In DSSAT, we simulate only monocropping production systems in single seasons (that is, cropping intensity = 1). By design, we decided not to include rotations or mixed systems (for example, rice-rice or rice-wheat) and to focus this book on the impacts of technologies on single crops. We track water and nitrogen balances with DSSAT, but this only reflects changes in use of resources. In other words, we do not imply changes in terms of environmental impacts (for example, water quality).

DSSAT uses a single assumption of 100 percent adoption to provide input data to IMPACT. In other words, if we consider a single pixel of 60 kilometers by 60 kilometers with areas cultivated in maize, we are comparing average yields obtained through adoption of technology X across the entire maize area in the pixel, with average yields obtained under baseline technologies/ practices across the same entire maize area of the pixel. We assume that farmers will adopt one of the simulated technologies only if yields are superior compared to the baseline yields (that is, for IMPACT, we assume that in the areas where DSSAT yields under the alternative technology are below the baseline crop yields, the baseline yields will be maintained). We call this the "smart farmer assumption."

Furthermore, in DSSAT, we assume that the assessed technologies are not yet adopted in the baseline, except for no-till in six countries. This assumption can potentially lead to the overestimation of yield impact for those technologies that have already been adopted to some extent in some countries. The relative share of current adoption is described in Chapter 2 in the literature review on agricultural technologies. Overall, the literature shows low adoption of almost all these technologies. For NUE and heat tolerance as specified here, there is no adoption to date. There is close to zero adoption for drought tolerance and little adoption of PA, drip irrigation, and sprinkler irrigation. Crop protection is defined with respect to the reduction of existing losses that occur under existing crop protection, so all benefits are incremental to the base value. However, we almost certainly overestimate returns from no-till, which is also widely adopted outside the six countries mentioned.

Our evaluation of the different technologies does not imply that they would be all similarly easy to develop, deploy, and adopt. Socioeconomic and technology levels, as well as economic and policy opportunities in different countries, will dictate the opportunity of adopting one or another technology. For this study, different rates of adoption and the shape of the adoption curve for each technology were assumed in IMPACT, but the rates of adoption for a given technology were assumed to be identical across regions. The adoption ceilings included a notion of socioeconomic factors affecting adoption, including cost, but with several additional limitations spelled out in the section on adoption profiles. Specifically, adoption ceilings were not differentiated by crop and region, differentiated regional or predominant farm characteristics were not reflected in the adoption pathway profiles assumed for this study, and the speed of adoption was kept constant across technologies. These assumptions limit the accuracy of the results.

No-till is modeled assuming continuous usage of no-till and minimum soil disturbance through 2050. Compared to conventional tillage, the continuous use of no-till results in the extensive build-up of carbon, organic matter, and soil moisture holding capacity that collectively lead to significant crop yield improvements over time. However, in practice, farmers frequently break the use of no-till by plowing their land for 1 or more years over a period of years. Such intermittent plowing substantially reduces the long-term yield impact of no-till. Therefore, our estimation of the long-run impact of no-till on crop yields may be too high relative to actual implementation in farmers' fields.

IMPACT simulations include endogenous price changes to equilibrate supply and demand, but the implications of these price changes for technology profitability—and hence for its rate of adoption—are not directly fed back into the adoption profiles. However, IMPACT captures the feedback effects from the adoption of the agricultural technologies. The initial shock from the adoption of the technology results in reduced prices, which lowers yields compared to the initial shocks. Thus, the final impact on yields (and area) is lower than the initial impact, thereby capturing the feedback effect of adoption. The initially reduced prices also induce higher demand, which limits the price impact, resulting in higher prices than the initial technology shock would predict and thus no overestimation of the price effect. Thus, the lower adoption rate is implicitly part of the yield decline effect.

The version of IMPACT that we currently use does not allow direct factoring in of the cost of development of agricultural technologies. A new version of IMPACT is under development, which will eventually include such analyses, but lack of globally consistent cost data will remain a challenge.

Any complex and multidimensional modeling effort, such as the present one, is obliged to make various simplifying assumptions, and even more assumptions and uncertainties confront models that project so far into the future. Consequently, simulated outcomes are not intended to be taken at face value but rather to demonstrate possible orders of magnitudes that suggest areas of focus that should be studied further as more data and more advanced analytical models are developed.

Assumptions about Crop Protection

As yields increase and production becomes more intensive, weeds, diseases, and insects will become increasing problems. But simulating the dynamic behavior of pest populations and their interactions with crops is a substantial project in its own right (Willocquet et al. 2002). DSSAT only allows the application of specific amounts of damage applied externally; that is, it does not represent the dynamics of pest populations. Developing good global models for the dynamics of the set of all important pests, and linking them effectively with socioeconomic models, is at the frontier of research in pest-crop interaction and is beyond the scope of the current book.

Crop protection is generally defined as any means that effectively reduces damages caused by pests and therefore benefits crop yields. To analyze the effects of crop protection on yields, we used data from reported chemical control of rice, wheat, and maize pests, which have been harmonized across different regions of the world (Oerke et al. 1994; Oerke and Dehne 2004; Oerke 2006). Comparable estimates of the effectiveness of alternative methods of pest control are not available across our study regions or at the global level.

These data are static and do not reflect emerging pests and diseases, ecological interactions between crops and pests, consequences of pesticide applications,[13] the rate of pest evolution and adaptation to both plant defenses and climate, and the misuse of pesticides.[14] However, our disaggregated estimates of pest prevalence that are based on regional-level damage data do take into

---

13 For example, brown plant hopper in Asia is a secondary problem stemming from insecticide spraying for leaf-feeding insects in the early crop stages (Way and Heong 1994; Bottrell and Schoenly 2012).

14 The effects of pesticides on ecosystem health, farm workers, nearby settlements, and consumer health are also beyond the goals of the present study.

consideration climatic factors specific to the two climate scenarios used in this study (Appendix C). Because of data limitations, we used a single species from each pest group (PAW) to represent that group and to approximate the prevalence of the group. These single pests are responsible for a substantial part of the damage across many regions. We furthermore assumed that, given the balance between pace of improvements in chemical effectiveness and evolution of the pests, these same pests may still represent the key pests in the future.

This aspect of the work has additional limitations that are directly linked to the current gaps in many areas of knowledge surrounding the evolution of crop-pest relationships under different climates. A summary of the limitations and uncertainty about the future distribution and impact of pests on crops is as follows:

1. We deal only with averages, but climate variability as well as changing climate averages can affect pests' life cycles and range.

2. We do not consider that new pests can emerge because of changing environmental conditions, or the effects of these conditions on plants' defenses, or the interaction of these factors and pests' life cycles with specific agricultural practices and technologies.

3. We do not consider how quickly pests can adapt to new climate conditions or how large the impacts of other factors (such as crop density) could be.

4. We do not consider the effects of pesticides on ecosystem health, farm workers, downstream settlements, and consumer health (although we do not take these issues lightly and are conscious of their importance across agriculture and public health).

Another limitation has to do with evolution of pest resistance to pesticides, which may appear at some point in the future for a specific type of chemical as a result of climate change and other factors. Just as resistance to pathogens and insect pests is an important trait of crop varieties, pathogens, insects, and weeds can develop resistance to specific pesticides, limiting the utility of a pesticide and requiring development of new pesticides. Even as yields increase and production becomes more intensive, weeds, diseases, and insects may increasingly become problems, especially as their ranges move to new areas. However, we do not have any reason to believe that innovation in future crop protection technology R&D would slow down in the face of potential pest evolution. Indeed, new R&D may effectively overcome such resistance. This is why we maintain the estimated magnitude of efficacy into the future. However, as mentioned above, we did hold the actual yield loss (that can be potentially avoided by applying crop protection technologies) constant over time. As such, if climate change makes pest epidemics and yield losses worse, then our estimates of the resultant crop protection will not be enhanced, because we apply the same effectiveness percentage. That is, in the face of potentially greater pest outbreaks and yield losses, the percentage effectiveness rates will generate higher levels of prevented damage. Since we do not do this, we are in fact underestimating the impact of crop protection by not considering the possible higher level of prevented expected pest damages in the future by continuing with current expected pest damage levels.

Considering the limitations on and the different sources of uncertainty about pest ranges and their interaction with climate, soil, and technology, the crop protection simulations reported in this book can only be considered as exploratory scenarios of yield change.

Chapter 4

DSSAT Results: Yield Impacts from the Process-Based Models

D SSAT simulations were carried out for maize-, rice-, and wheat-growing areas across the world on a 30-arcminute grid (approximately 60 kilometers by 60 kilometers) for two climate scenarios. The DSSAT analysis captures the biophysical and climatic variability across the world by using location-specific data on climate and soils, and simulated yields are the result of how our modeled technologies interact with these factors.

Table 4.1 illustrates how climate change simulated through the MIROC A1B and CSIRO A1B scenarios interacts with the agricultural practices encoded in the DSSAT baseline to affect baseline yields of maize, rice, and wheat by 2050, compared to yields in 2010. Climatic changes have negative effects for all crops and both rainfed and irrigated systems, but impacts are particularly strong for rice. For maize and rice, the largest adverse impacts are for rainfed systems, which account for most maize production, whereas most rice is irrigated. For wheat, negative impacts are largest in irrigated systems, which are concentrated in South Asia.

To estimate the biophysical yield changes over the baseline scenario that stem from the technologies studied, given soil and climate change scenarios,

TABLE 4.1 Effect of climate change on average maize, rice, and wheat yields, based on process-based models (DSSAT), between 2010 and 2050 (%)

	CSIRO A1B		MIROC A1B	
Crop	Rainfed	Irrigated	Rainfed	Irrigated
Maize	−13.2	−2.6	−16.5	−10.3
Rice	−23.2	−14.9	−24.8	−15.8
Wheat	−8.2	−9.3	−7.9	−10.8

Source: Authors.

Notes: Yields in 2010 are calculated as the average of the yields simulated during 2011–2020, and yields in 2050 are an average of the yields simulated during 2041–2050. A1B = greenhouse gas emissions scenario that assumes fast economic growth, a population that peaks mid-century, and the development of new and efficient technologies, along with a balanced use of energy sources; CSIRO = Commonwealth Scientific and Industrial Research Organisation's general circulation model; DSSAT = Decision Support System for Agrotechnology Transfer; MIROC = Model for Interdisciplinary Research on Climate.

we first simulated yields for the DSSAT baseline and then for each technology option over 40 years (2010–2050) for the three cereal crops. The following sections present yield impacts, calculated as the percentage difference of the yields with the technology over baseline yields. Results are aggregated at the country, region, and global levels, as well as across rainfed and irrigated areas using area-weighted average yields. The discussion mostly uses the MIROC A1B scenario results but highlights key differences for the CSIRO A1B scenario results.

Ex ante Yield Impacts: Global Results

Figure 4.1 presents the global aggregate yield impacts of the agricultural technologies under the MIROC and CSIRO climate scenarios.

Based on the DSSAT model results, under the hotter, wetter MIROC A1B climate scenario, the largest ex ante yield impacts are achieved with heat tolerance for maize, followed by no-till (see also Figures 4.2 and 4.3). Heat-tolerant varieties of maize show particularly high ex ante impacts, as they counteract the shortening of the crop duration under climate change, which otherwise would

FIGURE 4.1   Global yield impacts compared to the baseline scenario, by crop, MIROC A1B and CSIRO A1B scenarios, 2050 (%)

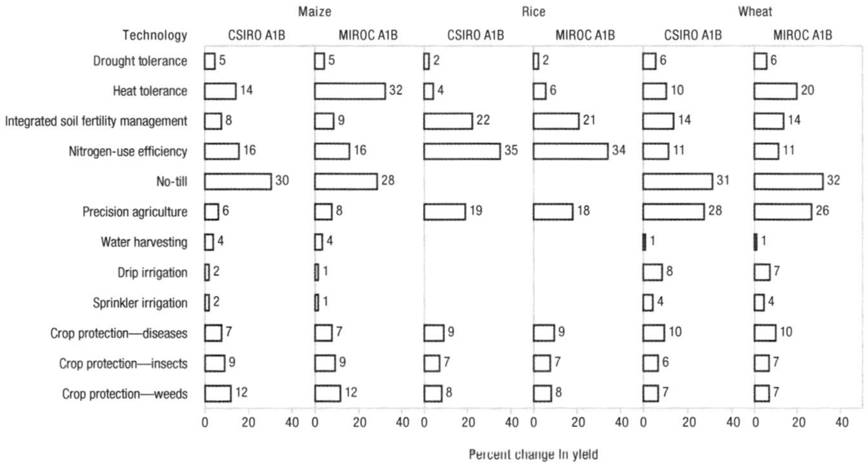

Technology	Maize CSIRO A1B	Maize MIROC A1B	Rice CSIRO A1B	Rice MIROC A1B	Wheat CSIRO A1B	Wheat MIROC A1B
Drought tolerance	5	5	2	2	6	6
Heat tolerance	14	32	4	6	10	20
Integrated soil fertility management	8	9	22	21	14	14
Nitrogen-use efficiency	16	16	35	34	11	11
No-till	30	28			31	32
Precision agriculture	6	8	19	18	28	26
Water harvesting	4	4			1	1
Drip irrigation	2	1			8	7
Sprinkler irrigation	2	1			4	4
Crop protection—diseases	7	7	9	9	10	10
Crop protection—insects	9	9	7	7	6	7
Crop protection—weeds	12	12	8	8	7	7

Percent change in yield

Source: Authors.

Notes: Yield impacts are not additive by technology. A1B = greenhouse gas emissions scenario that assumes fast economic growth, a population that peaks mid-century, and the development of new and efficient technologies, along with a balanced use of energy sources; CSIRO = Commonwealth Scientific and Industrial Research Organisation's general circulation model; MIROC = Model for Interdisciplinary Research on Climate.

FIGURE 4.2   Global map of yield impacts for rainfed maize, heat-tolerant varieties, compared to baseline scenario, MIROC A1B scenario, 2050 (%)

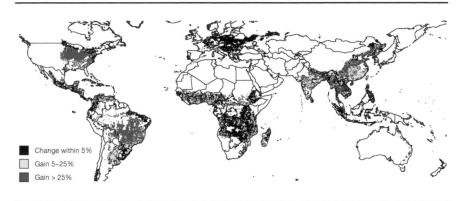

Source: Authors.
Note: A1B = greenhouse gas emissions scenario that assumes fast economic growth, a population that peaks mid-century, and the development of new and efficient technologies, along with a balanced use of energy sources; MIROC = Model for Interdisciplinary Research on Climate.

FIGURE 4.3   Global map of yield impacts for rainfed maize, no-till, compared to the baseline scenario, MIROC A1B scenario, 2050 (%)

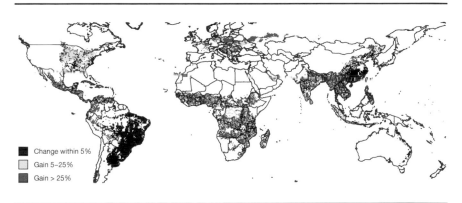

Source: Authors.
Notes: Change within 5 percent includes countries where no-till is not simulated as already adopted in the baseline. A1B = greenhouse gas emissions scenario that assumes fast economic growth, a population that peaks mid-century, and the development of new and efficient technologies, along with a balanced use of energy sources; MIROC = Model for Interdisciplinary Research on Climate.

adversely affect the grain filling of maize. Our no-till simulations assume long-term build-up of soil quality through continuous no-till over 40 years without interruption. No-till shows strong impacts, as drought is a major constraint for maize in key planting areas of the world and maize also heavily depends on nutrient inputs. No-till is therefore an ideal technology to improve outcomes in both

FIGURE 4.4   Global map of yield impacts for irrigated rice, nitrogen-use efficiency, compared to the baseline scenario, MIROC A1B scenario, 2050 (%)

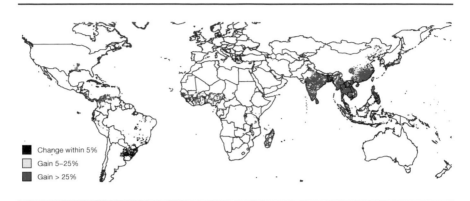

■ Change within 5%
☐ Gain 5–25%
■ Gain > 25%

Source: Authors.
Note: A1B = greenhouse gas emissions scenario that assumes fast economic growth, a population that peaks mid-century, and the development of new and efficient technologies, along with a balanced use of energy sources; MIROC = Model for Interdisciplinary Research on Climate.

areas. Moreover, maize is better able to respond to the technology when irrigated than when rainfed, because the irrigated crop does not experience water stress.

NUE has the highest yield impact for rice (Figures 4.1 and 4.4), followed by the combination of three types of crop protection (disease, insects, and weeds)[1] and ISFM. The NUE response in rice is due to generally higher nitrogen losses to leaching and volatilization for this technology in the baseline case. Agronomically, losses of nitrogen in flooded rice are high compared with either maize or wheat. Similarly, ISFM supports additional nutrient availability.

For wheat, no-till has the highest yield impact, followed by PA, and the sum of the three types of crop protection ranks third. Similarly to maize, no-till can address both nutrient and water shortages of wheat and thus respond to the highly negative yield impacts of irrigated wheat under climate change, as shown in Table 4.1. As described in Chapter 3 and Appendix A, the DSSAT baseline simulation reflects relatively low-input agriculture in much of the world, with low plant densities, and fertilizer is only applied once at planting time for wheat (contrary to maize, which is given a side dressing at least once during the crop cycle). Under PA, plant densities and fertilizer use will be optimum, which boosts yield results.

For crop protection, yield impacts are slightly larger for weeds in maize, and for diseases in rice and wheat (Figure 4.1).

---

1   Refer to Chapter 3 (especially the section on assumptions and limits of the study) for details about the limits regarding the implementation of crop protection.

FIGURE 4.5   Global yield impacts compared to the baseline scenario, by crop and cropping system, MIROC A1B scenario, 2050 (%)

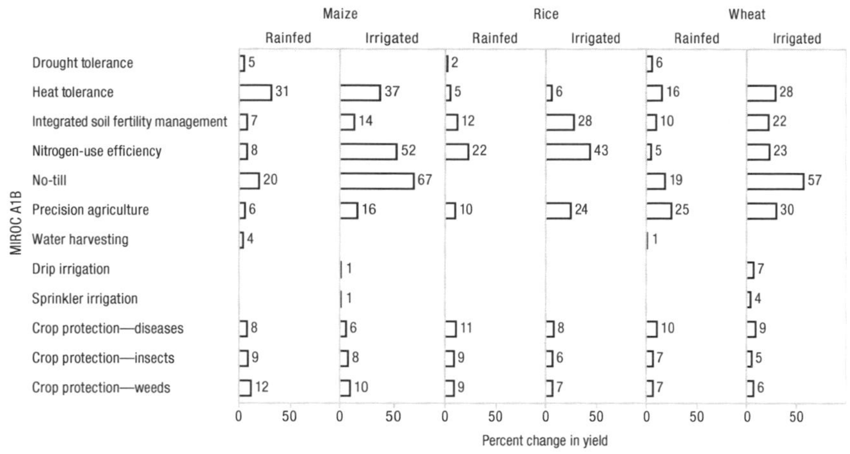

MIROC A1B

	Maize Rainfed	Maize Irrigated	Rice Rainfed	Rice Irrigated	Wheat Rainfed	Wheat Irrigated
Drought tolerance	5		2		6	
Heat tolerance	31	37	5	6	16	28
Integrated soil fertility management	7	14	12	28	10	22
Nitrogen-use efficiency	8	52	22	43	5	23
No-till	20	67			19	57
Precision agriculture	6	16	10	24	25	30
Water harvesting	4				1	
Drip irrigation		1				7
Sprinkler irrigation		1				4
Crop protection—diseases	8	6	11	8	10	9
Crop protection—insects	9	8	9	6	7	5
Crop protection—weeds	12	10	9	7	7	6

0   50   0   50   0   50   0   50   0   50   0   50

Percent change in yield

Source: Authors.

Note: A1B = greenhouse gas emissions scenario that assumes fast economic growth, a population that peaks mid-century, and the development of new and efficient technologies, along with a balanced use of energy sources; MIROC = Model for Interdisciplinary Research on Climate.

The ex ante yield impacts of switching from furrow to drip and sprinkler irrigation and of using water harvesting in rainfed conditions are not large, ranging from 1 to 7 percent depending on technology (Figure 4.5). This result is not surprising, given that cereal product quality is not particularly changed as a result of a switch from furrow to other types of irrigation and because water harvesting is a niche technology limited to those geographies where rainfall water capture is both sensible and yield enhancing. OA has no positive yield impacts. This result is due to the smart farmer assumption that only incorporates those geographic locations in the ex ante assessment where a technology outperforms the baseline technologies in the form of higher yields.

As Figure 4.5 shows, ex ante yield impacts are almost always higher under irrigated conditions. The differences are particularly striking for NUE and no-till in maize. As already explained, maize and the other cereals are better able to respond to the technology under irrigated conditions, because other yield-limiting factors (particularly, limited rainfall) cannot constrain the technologies evaluated. Thus, agricultural technology impacts are amplified with irrigation. Hence, continued investment in cost-effective irrigation should go hand in hand with technology rollout.

FIGURE 4.6   Global yield impacts compared to the baseline scenario, by crop and cropping system, combined technologies, MIROC A1B and CSIRO A1B scenarios, 2050 (%)

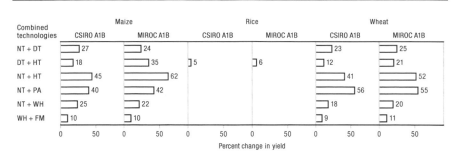

Source: Authors.
Note: A1B = greenhouse gas emissions scenario that assumes fast economic growth, a population that peaks mid-century, and the development of new and efficient technologies, along with a balanced use of energy sources; CSIRO = Commonwealth Scientific and Industrial Research Organisation's general circulation model; DT + HT = drought tolerance with heat tolerance; MIROC = Model for Interdisciplinary Research on Climate; NT + DT = no-till with drought tolerance; NT + HT = no-till with heat tolerance; NT + PA = no-till with precision agriculture; NT + WH = no-till with water harvesting; WH + FM = water harvesting with integrated soil fertility management.

In addition to the 11 individual technologies, 6 crop technology combinations were tested by applying two technologies simultaneously. Several of these combine enhanced crop management practices with advanced breeding strategies. Among these combined technologies, no-till with heat tolerance shows particular promise under the MIROC A1B climate scenario for maize and the combination of no-till with PA for wheat. In contrast, the combination of water harvesting with ISFM shows only limited promise (Figure 4.6). These results are in line with the ex ante yields of the individual technologies. It is understood that farmers might well choose other combinations or stack up to 11 of the technologies evaluated depending on the specific situation. We present these technology combinations as possible examples. Analyzing additional combinations would be too cumbersome and is only of interest for specific localities.

Figure 4.5 shows that ex ante yield impacts for drought tolerance[2] are relatively small globally—5 percent for maize, 2 percent for rice, and 6 percent for wheat across both climate change scenarios. However, more in-depth analysis shows that drought tolerance increases yields when most needed—during droughts—which is of key importance for the expected growing climate variability and extremes (Box 4.1).

---

2 Drought tolerance was parametrized as root enhancements to withstand water shortages and for maize also as reduced sensitivity of hybrid varieties to variability in ASI.

BOX 4.1 Drought tolerance

Drought tolerance is a desirable trait that allows agricultural producers to manage risk. It is of greatest concern during drought conditions. However, as a technology that influences risk, drought tolerance may not show large ex ante yield benefits when considering mean effects on productivity. Thus, for drought tolerance to be effective, performance under conditions at the drier end of the spectrum of reasonable weather for a particular location needs to be examined. Hence, to perform a meaningful assessment, we need a way to distinguish different intensities of drought and to match those up with the yield performance of the drought-tolerant technology. In this box, we briefly describe our findings for maize; a more complete exposition, which includes spring and winter wheat, can be found in Appendix A.

For this study, we define drought as a situation in which the crop would do better with irrigation. To implement this definition, we first use the automatic irrigation algorithms built into DSSAT to grow the crop in an idealized irrigated system that keeps a particular soil layer within a particular (high) range of soil moisture. At the end of the growing season, DSSAT reports how much cumulative irrigation was applied. We interpret this cumulative irrigation as an indication of total irrigation demand. The crop is then grown under a normal, rainfed situation using the same daily weather data. Finally, the drought-tolerant situation is modeled, again using the same daily weather. For this exercise, we use a random weather generator and repeat this process 567 times based on the monthly and annual climate means for the location. Each of the 567 years is completely independent of the others, and the initial conditions are reset for each year. The whole process results in 567 triplets of desired irrigation, normal yields, and drought-tolerant equivalent yields. The triplet results are then sorted in order by the cumulative irrigation water applied in the automatic irrigation case. The sorted list is divided into 21 bins or quantiles containing the results from 27 repetitions. Inside each bin, we compute the median values for the cumulative irrigation, normal yield, and drought-tolerant yield. This method allows us to see the general trend of the size of the benefit under different levels of local drought conditions.

For different locations, the greatest yield benefit could occur for different levels of (local) drought severity. Reasonable situations will be somewhere between extreme wetness and extreme droughts, with the greatest benefits being in the drier—but not necessarily the driest—conditions. To accommodate this diversity and still be able to assess the maximum reasonable benefit we might expect from drought tolerance, we also search pixel by pixel to determine which irrigation quantile is associated with the largest benefit from drought tolerance. Based on those maximum effect quantiles, we build equivalent maximum-effect yield maps, which allow us to visualize which locations could benefit most and to compute regional aggregations of these benefits.

We apply this approach to maize, a spring wheat variety, and a winter wheat variety on the appropriate production areas. Elsewhere in this book, several different varieties of these crops are used, depending on location. Rather than mix and match these varieties, for expositional purposes, we consider a single variety at a time. First, consider the geographic distribution of benefits. The maps in Box Figure 1 show the percentage improvement in yields when switching from the normal variety to the drought-tolerant version while keeping the same climatic conditions. The top map is for the quantile representing the lowest irrigation needs (least drought-like); the middle map is for the highest irrigation needs (most drought-like); the bottom map shows the highest benefit quantile by pixel. Results for winter and spring wheat can be found in Appendix A.

As expected, in the least drought-like conditions, drought tolerance pro-vides minimal benefits. Under the most drought-like conditions, drought tolerance provides a meaningful increase in yields. Moreover, there is a fair amount of diversity in how much benefit is possible across different locations even when selecting the maximum improvement in each location. The value of drought tolerance under drought conditions (and lack thereof otherwise) can be seen clearly in regional aggregations. Box Figure 2 takes the area-weighted average yields by quantile for China and the United States and shows the fractional improvement of the drought-tolerant variety over the equivalent normal variety. The improvement from the "maximum benefit" case is higher than that of any of the individual drought-intensity quantiles. However, as the maximum benefit is derived by considering all the differ-ent cumulative irrigation amounts, it does not have a single irrigation value that typifies it. To show it on the graph, we arbitrarily assign it a value of 500 millimeters to place the value to the far right of the actual quantile irriga-tion values.

We can examine the cumulative precipitation received by the plants in each quantile to assess how realistic the representation is and whether we are actually capturing droughts.

In the U.S. Corn Belt, a reasonable amount of rain for the season is about 450 millimeters (for example, http://www.extension.purdue.edu/extmedia/NCH/NCH-40.html). During extreme drought events like those of 1988 and 2012, rainfall amounts would drop 40–70 millimeters/month during June, July, and August for a total deficit from normal on the order of 100 millimeters. These values for normal and drought conditions corre-spond reasonably well with the middle and far-right ends of the quantiles

BOX FIGURE 1A   Drought impact maps for maize, baseline scenario, year 2000

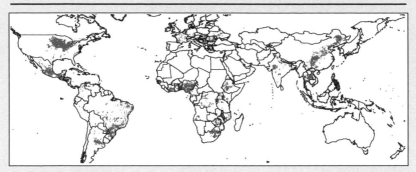

Least drought

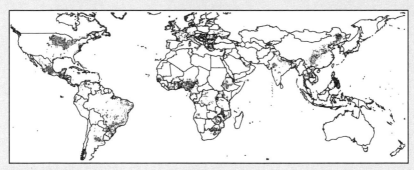

Most drought

Maximum benefit

▨ Old area lost	☐ Loss 3–5%	▨ Gain 5–10%
■ Loss > 10% of base	▨ Change within 3%	■ Gain > 10%
▨ Loss 5–10%	☐ Gain 3–5%	▨ New area gained

Source: Authors.

BOX FIGURE 1B   Drought impact maps for maize, CSIRO A1B scenario, year 2050

Least drought

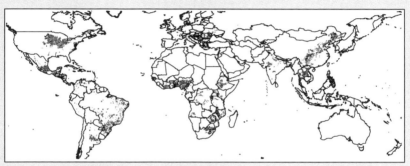

Most drought

Maximum benefit

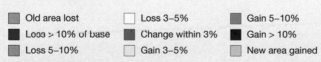

■ Old area lost	□ Loss 3–5%	■ Gain 5–10%
■ Loss > 10% of base	■ Change within 3%	■ Gain > 10%
■ Loss 5–10%	□ Gain 3–5%	■ New area gained

Source: Authors.

Note: A1B = greenhouse gas emissions scenario that assumes fast economic growth, a population that peaks mid-century, and the development of new and efficient technologies, along with a balanced use of energy sources; CSIRO = Commonwealth Scientific and Industrial Research Organisation's general circulation model.

BOX FIGURE 1C Drought impact maps for maize, MIROC A1B scenario, year 2050

Least drought

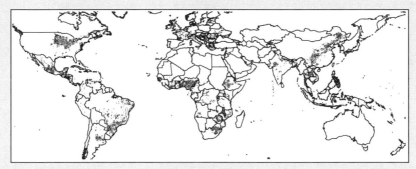

Most drought

Maximum benefit

Old area lost	Loss 3–5%	Gain 5–10%
Loss > 10% of base	Change within 3%	Gain > 10%
Loss 5–10%	Gain 3–5%	New area gained

Source: Authors.
Note: A1B = greenhouse gas emissions scenario that assumes fast economic growth, a population that peaks mid-century, and the development of new and efficient technologies, along with a balanced use of energy sources; MIROC = Model for Interdisciplinary Research on Climate.

BOX FIGURE 2   Ex ante yield benefits of drought tolerance compared to the original variety under three climate scenarios for China and the United States

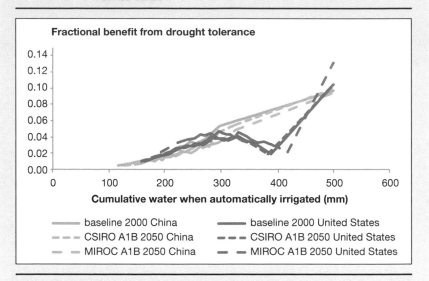

Source: Authors.
Note: A1B = greenhouse gas emissions scenario that assumes fast economic growth, a population that peaks mid-century, and the development of new and efficient technologies, along with a balanced use of energy sources; CSIRO = Commonwealth Scientific and Industrial Research Organisation's general circulation model; MIROC = Model for Interdisciplinary Research on Climate.

presented in Box Figure 3; that is, the total precipitation for the baseline climate in the middle quantiles is about 630 millimeters for the growing season and drops to 500 millimeters or less in the most drought-like quantiles.

In almost all cases, the drier conditions are associated with greater improvements from drought tolerance. In general, the amount of irrigation water farmers might desire for maize is higher under the future conditions than they are in the baseline, whereas for wheat the desired amounts do not change much. Because rainfed wheat tends to be grown in relatively cool conditions, this result is not surprising. However, not everything is similar. It is not clear whether the improvements under future conditions are usually larger than those under baseline conditions. Also, the potential benefits are different by country and crop.

Overall, we find that this particular implementation of a drought-tolerant trait for maize and wheat appears to behave as intended and

BOX FIGURE 3   Growing season precipitation by drought intensity compared to
the baseline scenario for maize in China and the United States,
2050 (mm)

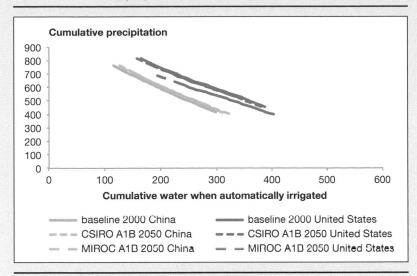

Source: Authors.
Note: A1B = greenhouse gas emissions scenario that assumes fast economic growth, a population that peaks
mid-century, and the development of new and efficient technologies, along with a balanced use of energy sources;
CSIRO = Commonwealth Scientific and Industrial Research Organisation's general circulation model; MIROC =
Model for Interdisciplinary Research on Climate.

shows meaningful benefits under drought conditions. The size of the benefit depends on the original variety the drought tolerance is incorporated into
and on local conditions. At the regional level, improvements in the neighbor-
hood of 9–13 percent could be achieved when drought conditions occur
under both current climate situations and those of the kind anticipated in
the future. Considering that the trait discussed here (roots with improved
access to soil moisture) is but a single mechanism, it is likely that a more
holistic bundle of traits would increase the envelope of yield improvements
under drought conditions to well above 10 percent. Finally, this sort of gridded crop simulation modeling can help map the diversity of outcomes,
because the benefits will not be uniform. Some locations will have substantially better outcomes (dark blue in the maps), whereas other show
limited impacts.

Source: Authors.

Ex ante Yield Impacts by Region

Figures 4.7–4.19 present the regional results of the various technologies, including OA. Ex ante impacts of agricultural technologies differ substantially by region and within regions by country. Across the three crops, the largest yield gains in percentage terms are in Africa, South Asia, and parts of Latin America and the Caribbean. Given the heterogeneity in yield response, it is important to target specific technologies to specific regions and countries. We find particularly high ex ante yield impacts for heat tolerance for North America and South Asia; drought tolerance for Latin America and the Caribbean, the Middle East and North Africa, and SSA; and crop protection for SSA, South Asia, and Eastern Europe and central Asia. PA shows the highest total gains for wheat and in major production areas in South Asia, the Middle East and North Africa and parts of Western Europe. NUE varieties are also critical to reduce resource use to further sustainable development and show gains in most developing regions, particularly in Latin America and the Caribbean and SSA. The largest potential for no-till is also in these two regions, whereas ISFM has benefits in low-input regions in Africa and parts of East Asia and the Pacific. In the following sections, we describe purely biophysical effects of these technologies, region by region. Of course, the actual introduction of such technologies in different regions will have to contend with differing investment costs and institutional and policy adjustments.

Africa South of the Sahara (SSA)

Among the three crops studied, maize is most important for SSA, but the importance of rice is growing. The DSSAT results indicate that no-till is the most yield-increasing technology for this region because of its soil-protection and water-enhancing properties under both climate change scenarios. Although maize is largely rainfed in the region at this point, irrigation development is growing rapidly, and both maize and rice (not assessed for no-till here) will increasingly benefit from irrigation. However, even rainfed maize sees a greater than 30 percent yield boost under no-till (Figure 4.7).

Improved NUE in maize and rice also shows largest benefits for SSA with more than 10 percent yield improvement by 2050 under rainfed conditions for both crops and up to 96 percent improvement for irrigated maize and a 50 percent yield increase for irrigated rice by 2050 under the CSIRO A1B scenario (Figure 4.15). This positive result again underlines the strong demand for enhanced nutrient—in particular, nitrogen—availability for cereal crops in the region.

FIGURE 4.7   Regional yield impacts compared to the baseline scenario, by crop and cropping system, no-till, MIROC A1B and CSIRO A1B scenarios, 2050 (%)

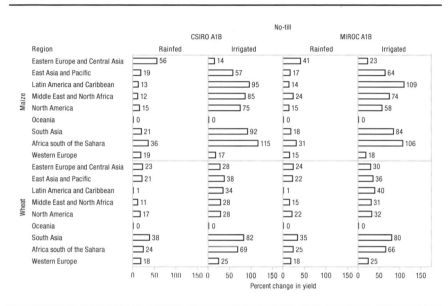

Source: Authors.
Note: A1B = greenhouse gas emissions scenario that assumes fast economic growth, a population that peaks mid-century, and the development of new and efficient technologies, along with a balanced use of energy sources; CSIRO = Commonwealth Scientific and Industrial Research Organisation's general circulation model; MIROC = Model for Interdisciplinary Research on Climate.

ISFM also shows large yield-enhancing benefits for maize in SSA compared to the DSSAT baseline scenario, with yields growing 21 percent under rainfed and 16 percent under irrigated conditions (Figure 4.8). Similar effects are seen for wheat, which is less common in the region, however. High ISFM impacts are likely due to the low levels of nutrients available in African soils, generally considered the key yield constraints in this region.

Moreover, drought tolerance shows major benefits in low rainfall environments of East Africa under the CSIRO A1B scenario (17 percent yield improvement) and still results in 7 percent improvement under the MIROC A1B scenario. Also, in higher-rainfall environments (rainfall greater than 500 millimeters per season), drought-tolerant crops do best in West and East Africa under both climate change scenarios (Figure 4.14).

Accelerated roll-out of crop protection for rainfed maize would have the largest ex ante yield impacts for SSA, with yield improvements in the range of 12–20 percent, depending on the cropping system and climate change scenario. For disease and insect control, only South Asia has similarly high yield benefits. Results for rainfed and irrigated rice in the SSA region are similarly high (Figures 4.16–4.18).

Among the combined technologies assessed, SSA shows high beneficial yield impacts of combined no-till and heat-tolerant varieties, with ex ante yield increases of more than 40 percent for rainfed and more than 100 percent for irrigated conditions under both climate change scenarios.

Asia

Rice remains the key staple crop grown in Asia, but wheat is also important in key Asian breadbaskets, particularly the Indo-Gangetic plains, and maize production is growing rapidly across the region as well.

SOUTH ASIA

Similar to SSA, South Asia sees large yield improvements from the alternative technologies assessed here. Yield gains are particularly high for no-till for both wheat and maize; for ISFM for rice and wheat; for PA for wheat; drought tolerance for wheat across all rainfall regimes; and NUE across all three cereals. South Asia also displays substantial benefits from advanced irrigation technologies for wheat, most likely due to the severe water shortages that the region already faces and that will be compounded as a result of climate change (Figure 4.11).

Heat tolerance is another technology with high potential in South Asia, particularly for maize and wheat. Irrigated maize yields are 66 percent higher with heat tolerance, and irrigated wheat yields are 33 percent higher under the MIROC climate change scenario. Yield improvements are lower but still substantial under the CSIRO climate change scenario (Figure 4.12).

Crop protection also results in higher yields ex ante, with largest benefits for maize through weed and insect control. In contrast, impacts for disease are roughly equally distributed across the three cereals, with yield improvements ranging from 1 to 33 percent (Figures 4.16–4.18).

Given that South Asia's wheat yields are under particular threat of adverse climate change effects, it is encouraging to see that a range of technologies can make major inroads in reducing these adverse effects for this key staple and breadbasket region.

EAST ASIA AND THE PACIFIC

Overall, ex ante yield improvements are lower in East Asia and the Pacific compared to the SSA and South Asia regions. No-till shows high potential in the region, with a 64 percent yield improvement for irrigated maize and a 36 percent yield improvement for irrigated wheat (Figure 4.7). ISFM shows

FIGURE 4.8 Regional yield impacts compared to the baseline scenario, by crop and cropping system, integrated soil fertility management, MIROC A1B and CSIRO A1B scenarios, 2050 (%)

| | | Integrated soil fertility management | | | |
| | | CSIRO A1B | | MIROC A1B | |
Crop	Region	Rainfed	Irrigated	Rainfed	Irrigated
Maize	Eastern Europe and Central Asia	2	0	2	0
	East Asia and Pacific	7	10	7	17
	Latin America and Caribbean	6	14	6	21
	Middle East and North Africa	15	10	30	16
	North America	0	6	0	7
	Oceania	1	7	1	6
	South Asia	4	9	3	8
	Africa south of the Sahara	21	16	21	16
	Western Europe	1	3	2	4
Rice	Eastern Europe and Central Asia		8	1	14
	East Asia and Pacific	12	31	10	31
	Latin America and Caribbean	6	25	6	24
	Middle East and North Africa	8	5	6	3
	North America		33		32
	South Asia	18	23	15	23
	Africa south of the Sahara	4	28	4	25
	Western Europe	114	47	27	23
Wheat	Eastern Europe and Central Asia	8	12	10	13
	East Asia and Pacific	13	23	14	18
	Latin America and Caribbean	8	15	9	15
	Middle East and North Africa	11	10	13	10
	North America	5	9	6	7
	Oceania	1	1	1	1
	South Asia	24	30	22	28
	Africa south of the Sahara	23	12	24	12
	Western Europe	11	12	12	11

Percent change in yield

Source: Authors.

Note: A1B = greenhouse gas emissions scenario that assumes fast economic growth, a population that peaks mid-century, and the development of new and efficient technologies, along with a balanced use of energy sources; CSIRO = Commonwealth Scientific and Industrial Research Organisation's general circulation model; MIROC = Model for Interdisciplinary Research on Climate.

large benefits for irrigated rice, with 31 percent higher yields compared to the DSSAT baseline and still 17 percent and 18 percent higher yields for irrigated maize and wheat, respectively. PA shows promise for all three irrigated cereals, with yield improvements of 26–29 percent (Figure 4.9).

NUE also results in large increases in yields for irrigated maize and rice in the region (Figure 4.15), indicating that nitrogen remains a constraint for some cropping systems and crops despite overall high fertilizer application rates.

Heat tolerance is another technology that needs further analysis in the East Asia and Pacific region, with potential yield impacts of 31 and 33 percent for rainfed and irrigated maize, respectively, and about 20 percent for wheat.

FIGURE 4.9 Regional yield impacts compared to the baseline scenario, by crop and cropping system, precision agriculture, MIROC A1B and CSIRO A1B scenarios, 2050 (%)

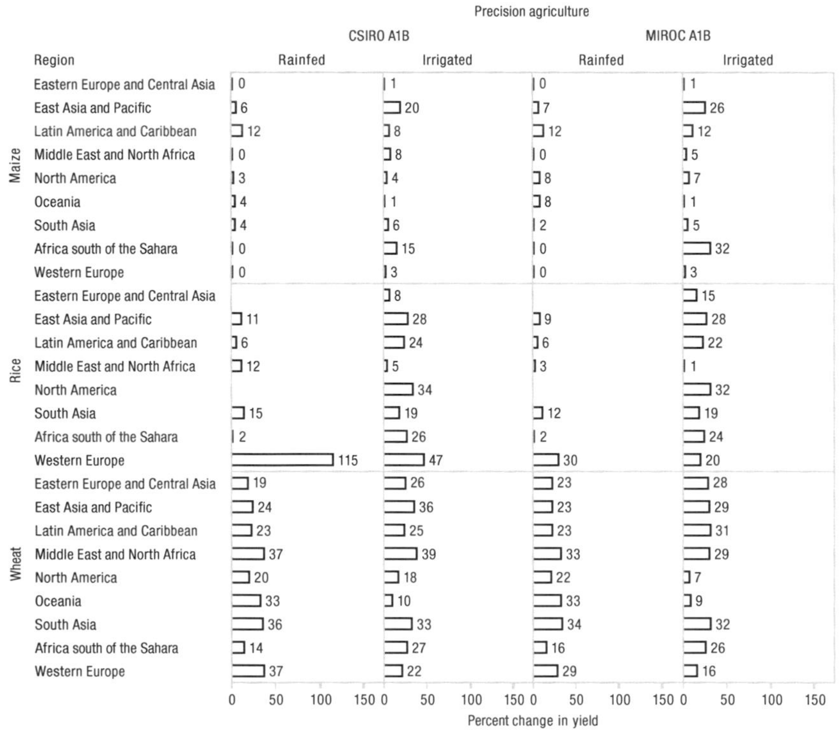

| | | Precision agriculture | | | |
| | | CSIRO A1B | | MIROC A1B | |
Crop	Region	Rainfed	Irrigated	Rainfed	Irrigated
Maize	Eastern Europe and Central Asia	0	1	0	1
	East Asia and Pacific	6	20	7	26
	Latin America and Caribbean	12	8	12	12
	Middle East and North Africa	0	8	0	5
	North America	3	4	8	7
	Oceania	4	1	8	1
	South Asia	4	6	2	5
	Africa south of the Sahara	0	15	0	32
	Western Europe	0	3	0	3
Rice	Eastern Europe and Central Asia		8		15
	East Asia and Pacific	11	28	9	28
	Latin America and Caribbean	6	24	6	22
	Middle East and North Africa	12	5	3	1
	North America		34		32
	South Asia	15	19	12	19
	Africa south of the Sahara	2	26	2	24
	Western Europe	115	47	30	20
Wheat	Eastern Europe and Central Asia	19	26	23	28
	East Asia and Pacific	24	36	23	29
	Latin America and Caribbean	23	25	23	31
	Middle East and North Africa	37	39	33	29
	North America	20	18	22	7
	Oceania	33	10	33	9
	South Asia	36	33	34	32
	Africa south of the Sahara	14	27	16	26
	Western Europe	37	22	29	16

Percent change in yield

Source: Authors.
Note: A1B = greenhouse gas emissions scenario that assumes fast economic growth, a population that peaks mid-century, and the development of new and efficient technologies, along with a balanced use of energy sources; CSIRO = Commonwealth Scientific and Industrial Research Organisation's general circulation model; MIROC = Model for Interdisciplinary Research on Climate.

Estimated crop protection benefits are more than 10 percent for weed control for maize (Figure 4.17) and 10 percent for insect control in rainfed maize (Figure 4.18).

Latin America and the Caribbean

Latin America and the Caribbean grows all three cereals studied here in different agroenvironments, with wheat concentrated in the southern part of the region and maize and rice dispersed across the region but concentrated in the center and in the Caribbean subregion. The potential to improve yields is

FIGURE 4.10 Regional yield impacts compared to the baseline scenario, by crop, water harvesting, MIROC A1B and CSIRO A1B scenarios, 2050 (%)

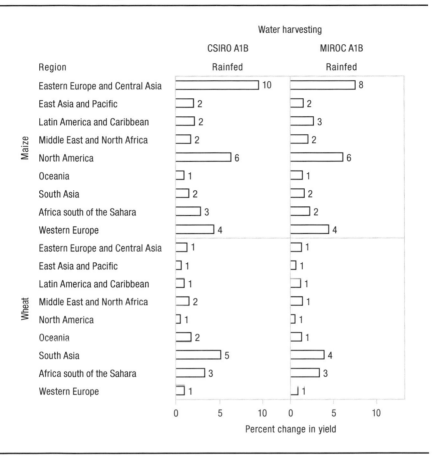

Source: Authors.
Note: A1B = greenhouse gas emissions scenario that assumes fast economic growth, a population that peaks mid-century, and the development of new and efficient technologies, along with a balanced use of energy sources; CSIRO = Commonwealth Scientific and Industrial Research Organisation's general circulation model; MIROC = Model for Interdisciplinary Research on Climate.

particularly high for irrigated maize, such as in Mexico, where water scarcity and adverse climate change impacts are rapidly increasing in importance.

The DSSAT ex ante yield assessments indicate that enhanced NUE has the highest yield impacts for irrigated maize in the region, up to 92 percent under the MIROC A1B scenario. The technology could improve irrigated rice yields by 26 percent and irrigated wheat yields by 13 percent (Figure 4.15). Similarly, the region shows the highest regional yield impacts for irrigated

FIGURE 4.11   Regional yield impacts compared to the baseline scenario, by crop and cropping system, advanced irrigation, MIROC A1B and CSIRO A1B scenarios, 2050 (%)

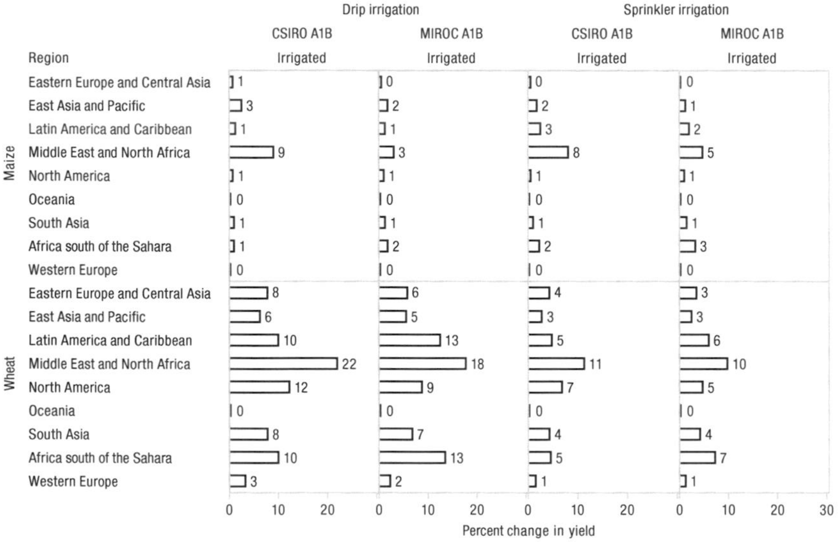

	Region	Drip irrigation		Sprinkler irrigation	
		CSIRO A1B Irrigated	MIROC A1B Irrigated	CSIRO A1B Irrigated	MIROC A1B Irrigated
Maize	Eastern Europe and Central Asia	1	0	0	0
	East Asia and Pacific	3	2	2	1
	Latin America and Caribbean	1	1	3	2
	Middle East and North Africa	9	3	8	5
	North America	1	1	1	1
	Oceania	0	0	0	0
	South Asia	1	1	1	1
	Africa south of the Sahara	1	2	2	3
	Western Europe	0	0	0	0
Wheat	Eastern Europe and Central Asia	8	6	4	3
	East Asia and Pacific	6	5	3	3
	Latin America and Caribbean	10	13	5	6
	Middle East and North Africa	22	18	11	10
	North America	12	9	7	5
	Oceania	0	0	0	0
	South Asia	8	7	4	4
	Africa south of the Sahara	10	13	5	7
	Western Europe	3	2	1	1

Percent change in yield

Source: Authors.

Note: A1B = greenhouse gas emissions scenario that assumes fast economic growth, a population that peaks mid-century, and the development of new and efficient technologies, along with a balanced use of energy sources; CSIRO = Commonwealth Scientific and Industrial Research Organisation's general circulation model; MIROC = Model for Interdisciplinary Research on Climate.

maize under no-till, up to 109 percent (Figure 4.7). ISFM could improve irrigated maize and rice yields by 21 and 24 percent, respectively (Figure 4.8). Furthermore, heat tolerance benefits could be large for irrigated wheat and rainfed maize (both at 31 percent).

For crop protection, potential yield benefits are substantial for enhanced weed control on maize and rice (Figure 4.17), and disease and insect control for rice (Figures 4.16 and 4.18).

Middle East and North Africa

The Middle East and North Africa region grows most crops in irrigated environments. Among the three staple crops assessed, it grows chiefly wheat and rice. Similar to other developing-country regions, no-till also has considerable promise in the region, particularly for irrigated maize (74 percent yield improvement), but also for wheat (31 percent yield improvement) (Figure 4.7). ISFM increases irrigated yields of maize by 16 percent and of

FIGURE 4.12   Regional yield impacts compared to the baseline scenario, by crop and cropping system, heat tolerance, MIROC A1B and CSIRO A1B scenarios, 2050 (%)

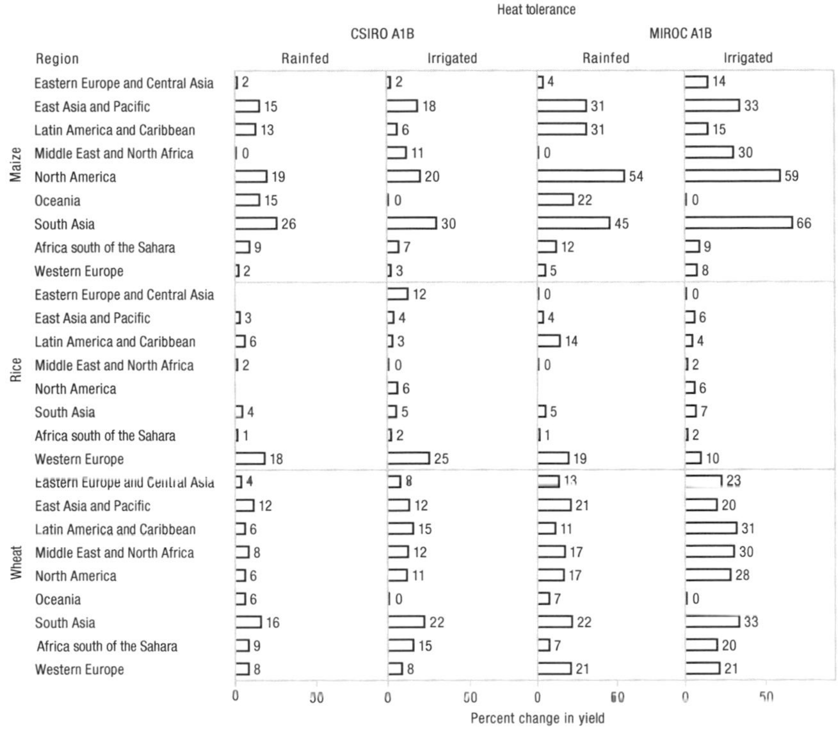

Source: Authors.

Note: A1B = greenhouse gas emissions scenario that assumes fast economic growth, a population that peaks mid-century, and the development of new and efficient technologies, along with a balanced use of energy sources; CSIRO = Commonwealth Scientific and Industrial Research Organisation's general circulation model; MIROC = Model for Interdisciplinary Research on Climate.

wheat by 10 percent (Figure 4.8). PA does particularly well for irrigated wheat, for the same reasons as described in the global yield assessment section. Yield improvements of 29–39 percent are simulated under the two climate change scenarios (Figure 4.9). The Middle East and North Africa is the region with the highest potential yield benefits from switching from furrow to drip or sprinkler irrigation for maize and wheat, with wheat yield benefits of 10–22 percent depending on scenario (Figure 4.11). The region is already quite hot today, and further temperature increases under climate change will particularly affect its wheat-growing areas. The heat-tolerant varieties assessed here show ex ante yield improvement potential of between 12 and 30 percent

FIGURE 4.13   Regional yield impacts compared to the baseline scenario, by crop and cropping system, drought tolerance, MIROC A1B and CSIRO A1B scenarios, 2050 (%)

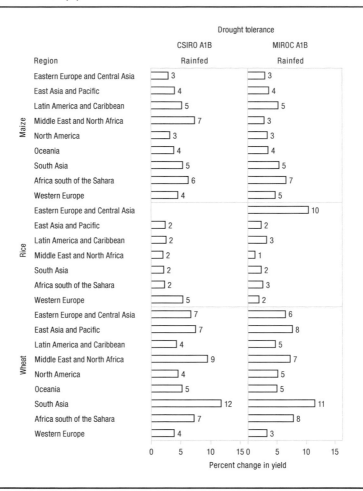

Source: Authors.
Note: A1B = greenhouse gas emissions scenario that assumes fast economic growth, a population that peaks mid-century, and the development of new and efficient technologies, along with a balanced use of energy sources; CSIRO = Commonwealth Scientific and Industrial Research Organisation's general circulation model; MIROC = Model for Interdisciplinary Research on Climate.

for irrigated wheat, depending on the climate change scenario, with the potential larger under the hotter and wetter MIROC A1B scenario. The region also has great potential for drought tolerance in wheat of 7–9 percent, depending on the climate change scenario (Figure 4.13). For maize, the region has some

FIGURE 4.14   Regional yield impacts compared to the baseline scenario, by crop and rainfall patterns, drought tolerance, MIROC A1B and CSIRO A1B scenarios, 2050 (%)

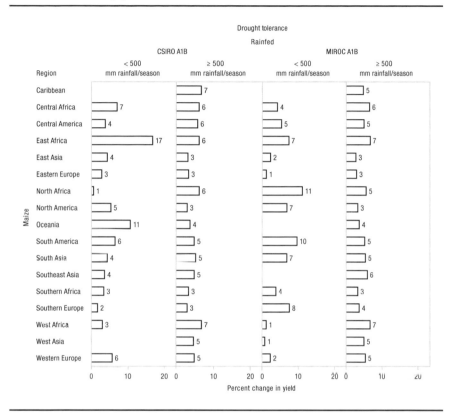

Source: Authors.
Note: A1B = greenhouse gas emissions scenario that assumes fast economic growth, a population that peaks mid-century, and the development of new and efficient technologies, along with a balanced use of energy sources; CSIRO = Commonwealth Scientific and Industrial Research Organisation's general circulation model; MIROC = Model for Interdisciplinary Research on Climate.

potential for drought tolerance in North Africa under low rainfall conditions, but the area benefiting would be small (Figure 4.14).

Similar to other regions, the Middle East and North Africa's irrigated maize yields are also likely to benefit from enhanced NUE in irrigated maize, with up to 57 percent yield improvement, but yield benefits would be less than 10 percent for rice and wheat (Figure 4.15).

Crop protection benefits for wheat areas in the Middle East and North Africa would be around 10 percent for disease control, and at the same level

FIGURE 4.15   Regional yield impacts compared to the baseline scenario, by crop and cropping system, nitrogen-use efficiency, MIROC A1B and CSIRO A1B scenarios, 2050 (%)

		CSIRO A1B		MIROC A1B	
	Region	Rainfed	Irrigated	Rainfed	Irrigated
Maize	Eastern Europe and Central Asia	7	9	7	14
	East Asia and Pacific	7	44	8	51
	Latin America and Caribbean	9	86	9	92
	Middle East and North Africa	8	49	14	57
	North America	2	56	1	46
	Oceania	0	1	0	1
	South Asia	16	66	13	48
	Africa south of the Sahara	16	96	15	89
	Western Europe	4	11	5	14
Rice	Eastern Europe and Central Asia		18	0	22
	East Asia and Pacific	22	43	22	43
	Latin America and Caribbean	9	28	9	26
	Middle East and North Africa	11	7	12	7
	North America		40		31
	South Asia	31	43	26	44
	Africa south of the Sahara	11	50	11	47
	Western Europe	25	28	20	33
Wheat	Eastern Europe and Central Asia	4	12	5	14
	East Asia and Pacific	9	22	10	20
	Latin America and Caribbean	1	13	1	13
	Middle East and North Africa	2	9	3	9
	North America	4	10	5	8
	Oceania	0	0	0	0
	South Asia	13	30	13	29
	Africa south of the Sahara	6	18	6	18
	Western Europe	5	10	6	11

Nitrogen-use efficiency

Percent change in yield

Source: Authors.

Note: A1B = greenhouse gas emissions scenario that assumes fast economic growth, a population that peaks mid-century, and the development of new and efficient technologies, along with a balanced use of energy sources; CSIRO = Commonwealth Scientific and Industrial Research Organisation's general circulation model; MIROC = Model for Interdisciplinary Research on Climate.

for insect control in maize. Crop protection benefits would be slightly lower for the other crops and types of crop protection (Figures 4.16–4.18).

North America

Rainfed maize is arguably the key staple crop of interest in North America, but wheat also matters for Canada, and substantial irrigated rice can be found in the United States. Both no-till and NUE improvements have large positive yield benefits for irrigated maize in North America, with increases ranging from 58 to 75 percent and from 46 to 56 percent, respectively. Corresponding rainfed yield gains are much lower at 15 percent and 1–2 percent, respectively

FIGURE 4.16   Regional yield impacts compared to the baseline scenario, by crop and cropping system, crop protection—diseases, MIROC A1B and CSIRO A1B scenarios, 2050 (%)

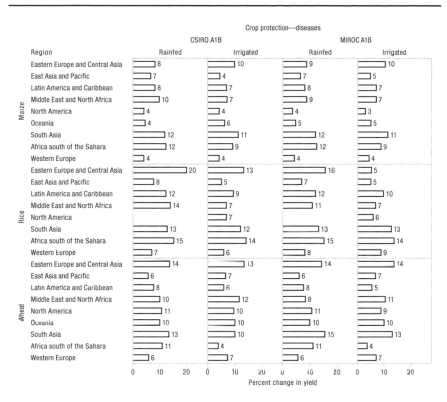

Source: Authors.

Note: A1B = greenhouse gas emissions scenario that assumes fast economic growth, a population that peaks mid-century, and the development of new and efficient technologies, along with a balanced use of energy sources; CSIRO = Commonwealth Scientific and Industrial Research Organisation's general circulation model; MIROC = Model for Interdisciplinary Research on Climate.

(Figures 4.7 and 4.15). No-till is also yield-improving for both rainfed and irrigated wheat (17–32 percent, depending on scenario); NUE has large potential yield gains for irrigated rice (31–40 percent, depending on scenario).

Given the substantial temperature increases expected under climate change in parts of cereal-growing North America, we find large ex ante yield improvement potential for heat-tolerant varieties (Figure 4.12), particularly under the hotter MIROC A1B climate change scenario. Under this scenario, yields could improve by 54 and 59 percent for rainfed and irrigated maize, respectively; yield impacts would be 19–20 percent under the CSIRO climate change scenario. Heat tolerance also has important yield effects for wheat of

FIGURE 4.17   Regional yield impacts compared to the baseline scenario, by crop and cropping system, crop protection—weeds, MIROC A1B and CSIRO A1B scenarios, 2050 (%)

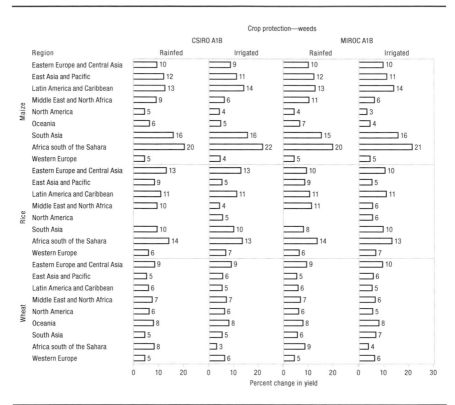

Source: Authors.

Note: A1B = greenhouse gas emissions scenario that assumes fast economic growth, a population that peaks mid-century, and the development of new and efficient technologies, along with a balanced use of energy sources; CSIRO = Commonwealth Scientific and Industrial Research Organisation's general circulation model; MIROC = Model for Interdisciplinary Research on Climate.

6–28 percent, depending on cropping system and climate change scenario. The potential for drought-tolerant varieties across the region is 5–7 percent in low-rainfall areas (Figure 4.14). More details on the potential for drought tolerance for the United States are presented in Box 4.1.

PA shows high potential yield benefits for irrigated rice (32–34 percent, depending on climate change scenario) and also substantial benefits for rainfed wheat (Figure 4.9). Most benefits from crop protection are already reaped in North America, but some potential exists for disease control for wheat (Figure 4.16).

FIGURE 4.18   Regional yield impacts compared to the baseline scenario, by crop and cropping system, crop protection—insects, MIROC A1B and CSIRO A1B scenarios, 2050 (%)

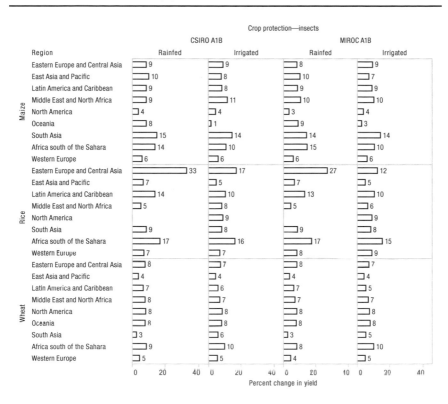

Source: Authors.

Note: A1B = greenhouse gas emissions scenario that assumes fast economic growth, a population that peaks mid-century, and the development of new and efficient technologies, along with a balanced use of energy sources; CSIRO = Commonwealth Scientific and Industrial Research Organisation's general circulation model; MIROC = Model for Interdisciplinary Research on Climate.

Europe and Central Asia

WESTERN EUROPE

No-till shows promise for both maize and wheat in Western Europe, with yield increases ranging from 15 to 19 percent and from 18 to 25 percent, depending on cropping system and climate change scenario (Figure 4.7). ISFM shows high potential for rice production (which is a minor crop in the region) and also substantial yield benefits for wheat, ranging from 11 to 12 percent (Figure 4.8). Similarly, PA shows high potential for rice and also for wheat (Figure 4.9). Wheat yields might grow by 16–37 percent with PA tools, depending on cropping system and climate change scenario. NUE can improve irrigated maize yields in Western Europe by more than 14 percent, irrigated rice yields by up to 33 percent, and irrigated wheat yields by 11 percent (Figure 4.15). Similar to the North America region, additional yield benefits from full adoption of crop protection would be small.

EASTERN EUROPE AND CENTRAL ASIA

Wheat is a key crop in this region, particularly in Central Asia, which is plagued by substantial water shortages, environmental degradation, and considerable but highly uncertain impacts from climate change.

No-till for maize and wheat as well as PA for wheat show high ex ante yield impacts in Eastern Europe and Central Asia. No-till yield benefits for maize range from 14 to 56 percent, depending on the cropping system and climate change scenario; no-till yield benefits for wheat would be 23–30 percent (Figure 4.7). PA could improve rainfed and irrigated wheat yields by 19–28 percent (Figure 4.9).

Drip irrigation shows some promise for wheat, with yield gains of 6–8 percent (Figure 4.11). Water harvesting shows yield improvements for maize in the region (8–10 percent) (Figure 4.10). Heat tolerance shows potential benefits under the MIROC A1B scenario for wheat (13–23 percent improvement).

Crop protection has the potential to improve yields in the region, particularly for disease control in wheat (13–14 percent) and for disease and insect control in rainfed rice. Values are just below 10 percent for the remaining crop protection elements and crops assessed (Figures 4.16–4.18).

Results for Organic Agriculture

Figure 4.19 presents simulation results for OA based on data from Seufert, Ramankutty, and Foley (2012). The figure shows consistently decreased yields across regions and crops, with small fluctuations around the mean. Yield impacts are most negative for wheat.

The literature review and extensive consultations we conducted with experts in Brazil and India suggest that OA is unlikely to play a significant role in the technology mix for addressing food security at the global level.

Results for Resource Use

To balance considerations of productivity effects with those regarding resource-use efficiency, we explore global average levels of nitrogen losses for

FIGURE 4.19 Regional yield impacts by crop and cropping system, organic agriculture, MIROC A1B and CSIRO A1B scenarios, 2050 (%)

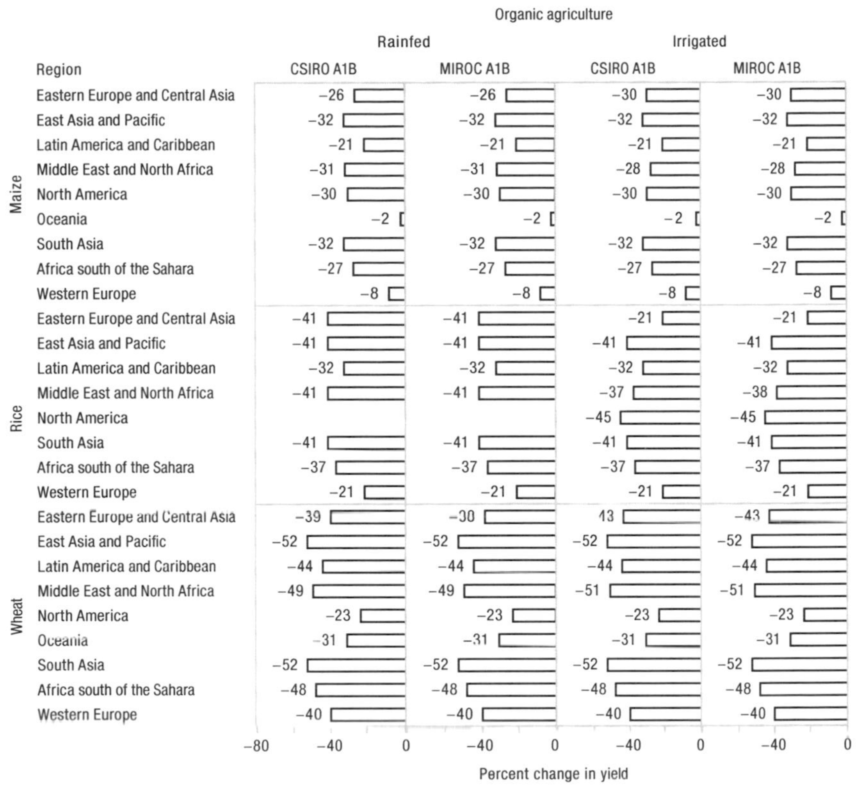

Organic agriculture

		Rainfed		Irrigated	
	Region	CSIRO A1B	MIROC A1B	CSIRO A1B	MIROC A1B
Maize	Eastern Europe and Central Asia	-26	-26	-30	-30
	East Asia and Pacific	-32	-32	-32	-32
	Latin America and Caribbean	-21	-21	-21	-21
	Middle East and North Africa	-31	-31	-28	-28
	North America	-30	-30	-30	-30
	Oceania	-2	-2	-2	-2
	South Asia	-32	-32	-32	-32
	Africa south of the Sahara	-27	-27	-27	-27
	Western Europe	-8	-8	-8	-8
Rice	Eastern Europe and Central Asia	-41	-41	-21	-21
	East Asia and Pacific	-41	-41	-41	-41
	Latin America and Caribbean	-32	-32	-32	-32
	Middle East and North Africa	-41	-41	-37	-38
	North America			-45	-45
	South Asia	-41	-41	-41	-41
	Africa south of the Sahara	-37	-37	-37	-37
	Western Europe	-21	-21	-21	-21
Wheat	Eastern Europe and Central Asia	-39	-30	43	-43
	East Asia and Pacific	-52	-52	-52	-52
	Latin America and Caribbean	-44	-44	-44	-44
	Middle East and North Africa	-49	-49	-51	-51
	North America	-23	-23	-23	-23
	Oceania	-31	-31	-31	-31
	South Asia	-52	-52	-52	-52
	Africa south of the Sahara	-48	-48	-48	-48
	Western Europe	-40	-40	-40	-40

Percent change in yield

Source: Authors.
Note: A1B = greenhouse gas emissions scenario that assumes fast economic growth, a population that peaks mid-century, and the development of new and efficient technologies, along with a balanced use of energy sources; CSIRO = Commonwealth Scientific and Industrial Research Organisation's general circulation model; MIROC = Model for Interdisciplinary Research on Climate.

each technology compared to the baseline, in 2050, under the two climate change scenarios. Based on our simulations and the specific representation of each technology in the crop model, no-till, NUE, and heat tolerance are the most promising technologies for reducing nitrogen leaching for maize. NUE, ISFM, and PA are the technologies with the largest reductions in nitrogen loss for rice; and no-till, ISFM, and PA are the key technologies with reductions in nitrogen losses for wheat (Figure 4.20). We also find small reductions in nitrogen losses for drought-tolerant varieties for maize and wheat and for drip and

FIGURE 4.20   Differences in nitrogen losses and nitrogen productivity compared to the baseline scenario, by crop and cropping system, global average, MIROC A1B and CSIRO A1B scenarios, 2050 (%)

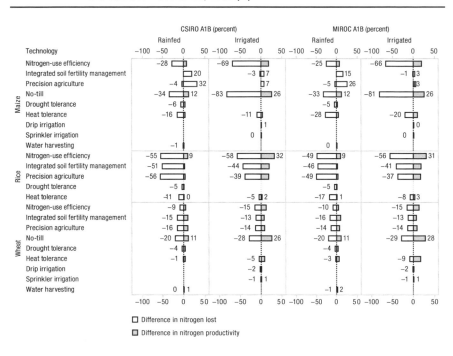

Source: Authors.

Notes: Negative numbers indicate reduced leaching compared to baseline. A1B = greenhouse gas emissions scenario that assumes fast economic growth, a population that peaks mid-century, and the development of new and efficient technologies, along with a balanced use of energy sources; CSIRO = Commonwealth Scientific and Industrial Research Organisation's general circulation model; MIROC = Model for Interdisciplinary Research on Climate.

sprinkler irrigation for wheat. The increased resource use efficiency of these technologies has complementary benefits for reductions of nitrogen loss.

Surprisingly, we find that ISFM and PA slightly increase nitrogen losses in rainfed maize. This is likely due to the higher volatilization of nitrogen for these technologies under rainfed conditions.

Similarly, we assessed changes in irrigation water use under the key technologies of drip and sprinkler irrigation. The results are presented in Figure 4.21. As expected, global water savings are larger for drip than for sprinkler irrigation. Irrigation water savings on field using drip irrigation are 24–27 percent, depending on crop and climate change scenario, whereas water savings for sprinklers are 11–12 percent.

FIGURE 4.21 Differences in irrigation water use and water productivity compared to the baseline scenario, by crop, global average, MIROC A1B and CSIRO A1B scenarios, 2050 (%)

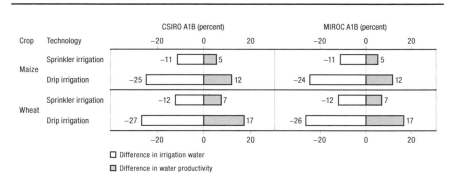

Source: Authors.

Note: A1B = greenhouse gas emissions scenario that assumes fast economic growth, a population that peaks mid-century, and the development of new and efficient technologies, along with a balanced use of energy sources; CSIRO = Commonwealth Scientific and Industrial Research Organisation's general circulation model; MIROC = Model for Interdisciplinary Research on Climate.

Chapter 5

IMPACT Results: Effects on Yields, Prices, Trade, and Food Security

A s described in detail in Chapter 3, we aggregate the biophysical productivity changes simulated in DSSAT at the pixel level to the FPUs and incorporate them in IMPACT using prespecified adoption profiles (Rosegrant and IMPACT Development Team 2012). By means of this integration, we simulate the interaction between changes in biophysical factors and population and economic growth and we estimate the resulting changes in food and water supply and demand, trade, and prices over four decades for the 11 technologies.

IMPACT Baseline Results

IMPACT baseline projections indicate substantial increases in world prices of maize, rice, and wheat between 2010 and 2050, with the price of maize almost doubling under MIROC A1B (Table 5.1). Tables 5.2 and 5.3 present changes in yields, production, area, and malnutrition levels under the baseline for the same period. Despite continued growth in cereal yields, the number of people at risk of hunger increases by about 10 percent between 2010 and 2050 across developing countries in both climate change scenarios. In SSA, the projected increase is a staggering 45–51 percent, and the share of malnourished children would grow by 4–5 percent as well (Table 5.3).

TABLE 5.1 Change in global prices of maize, rice, and wheat, between 2010 and 2050 (%)

Crop	MIROC A1B	CSIRO A1B
Maize	92	78
Rice	66	64
Wheat	73	65

Source: Authors.

Note: A1B = greenhouse gas emissions scenario that assumes fast economic growth, a population that peaks mid-century, and the development of new and efficient technologies, along with a balanced use of energy sources; CSIRO = Commonwealth Scientific and Industrial Research Organisation's general circulation model; IFPRI = International Food Policy Research Institute; IMPACT = International Model for Policy Analysis of Agricultural Commodities and Trade; MIROC = Model for Interdisciplinary Research on Climate.

TABLE 5.2   Change in production, yields, and harvested area, IMPACT baseline, MIROC A1B
and CSIRO A1B scenarios, selected regions, between 2010 and 2050 (%)

	MIROC A1B			CSIRO A1B		
Region	Maize	Rice	Wheat	Maize	Rice	Wheat
Production						
Developed	39.8	11.6	21.1	58.1	10.9	25.7
Developing	86.3	10.1	40.5	80.1	10.3	43.0
North America	40.0	39.7	40.9	63.5	45.5	47.9
South Asia	154.2	24.5	4.7	135.9	20.9	0.1
Africa south of the Sahara	76.0	187.2	119.1	61.2	189.1	120.3
Yields						
Developed	13.8	28.5	21.7	33.6	30.2	29.7
Developing	55.6	20.9	29.8	57.2	21.4	34.5
North America	11.4	27.9	41.7	33.4	34.0	52.8
South Asia	93.4	37.2	7.6	92.0	33.5	4.4
Africa south of the Sahara	60.3	93.6	56.2	52.2	95.6	60.8
Area harvested						
Developed	22.9	−13.1	−0.5	18.4	−14.8	−3.1
Developing	19.7	−8.9	8.2	14.6	−9.1	6.3
North America	25.7	9.2	−0.6	22.5	8.6	−3.2
South Asia	31.5	−9.3	−2.7	22.9	−9.4	−4.1
Africa south of the Sahara	9.8	48.3	40.3	5.9	47.8	37.0

Source: Authors.

Note: A1B = greenhouse gas emissions scenario that assumes fast economic growth, a population that peaks mid-century, and the development of new and efficient technologies, along with a balanced use of energy sources; CSIRO = Commonwealth Scientific and Industrial Research Organisation's general circulation model; IMPACT = International Model for Policy Analysis of Agricultural Commodities and Trade; MIROC = Model for Interdisciplinary Research on Climate.

TABLE 5.3   Change in hunger indicators, IMPACT baseline, selected regions, between 2010
and 2050 (%)

	MIROC A1B		CSIRO A1B	
Region	Number of people at risk of hunger	Malnourished children	Number of people at risk of hunger	Malnourished children
Developing	11.4	−22.6	9.8	−23.3
South Asia	3.8	−27.5	−0.7	−27.9
Africa south of the Sahara	50.8	4.7	44.7	3.6

Source: Authors.

Note: A1B = greenhouse gas emissions scenario that assumes fast economic growth, a population that peaks mid-century, and the development of new and efficient technologies, along with a balanced use of energy sources; CSIRO = Commonwealth Scientific and Industrial Research Organisation's general circulation model; IMPACT = International Model for Policy Analysis of Agricultural Commodities and Trade; MIROC = Model for Interdisciplinary Research on Climate.

Results for Alternative Technologies:
MIROC A1B Climate Change Scenario

Changes in Cereal Prices under Alternative Technologies

As exemplified by the recent world food-price crises, shifts in prices are the result of a combination of complex interlinking factors. In 2008, climate shocks, increased energy costs, growing biofuel demand, and other economic factors adversely affected the production of food, triggering spikes in food prices. These spikes led to reduced access to food for the poor and resulted in the introduction of many trade-distorting measures that pushed up food prices even further (Headey and Fan 2010; Fan, Torero, and Heady 2011).[1] When focusing on long-term trends, large sustained shifts in world prices for major commodities (like rice, maize, and wheat) signal substantial changes and imbalances in supply and demand; some of the most common causes of these imbalances are higher demand stemming from population and income growth, declines in yields or production due to climate change, or both (Nelson et al. 2010).

The IMPACT baseline projects an increase in world prices of rice, wheat, and maize under both climate scenarios. In 2050, all our scenarios of technology adoption show a decline in prices compared to the baseline. Under both MIROC A1B and CSIRO A1B, no-till and heat-tolerant varieties have the strongest price-reduction effects for maize and wheat, compared to baseline prices in 2050. NUE varieties and PA have the strongest potential to reduce prices for rice (Table 5.4). Moreover, among the combined technologies, heat-tolerant varieties grown under no-till and PA with no-till achieve reductions in prices of between 10 and 20 percent for maize and wheat compared to the baseline.

To better understand the source of these changes in prices, we look at the global effect of technology options on yields, production, trade, and food security outcomes. We then explore the regional impacts of technologies by focusing on a group of selected regions that includes some of the breadbaskets of the world (for example, North America) and on some of the regions most at risk of malnutrition (for example, SSA). We base our description on the MIROC A1B climate scenario and then comment on possible changes under CSIRO A1B.

Changes in Cereal Yields under Alternative Technologies

At the global level, NUE, no-till, heat tolerance, and PA show the largest increases in yields compared to the baseline in 2050. Heat-tolerant varieties

---

1 http://www.ifpri.org/pressrelease/study-challenges-conventional-wisdom-causes-global-food-crisis-recommends-reforms-preve.

TABLE 5.4   Change in world prices of wheat, rice, and maize compared to the baseline scenario, by technology, 2050 (%)

	MIROC A1B			CSIRO A1B		
Technology	Wheat	Rice	Maize	Wheat	Rice	Maize
No-till	−14.8	−2.7	−15.5	−14.2	−2.7	−15.2
Nitrogen-use efficiency	−8.4	−20.3	−12.0	−8.2	−20.4	−11.1
Heat tolerance	−9.7	−5.8	−15.5	−5.4	−3.6	−7.6
Precision agriculture	−9.7	−10.3	−4.9	−10.6	−10.5	−3.7
Crop protection—weeds	−2.8	−3.1	−3.2	−2.9	−3.2	−3.0
Crop protection—insects	−2.7	−3.1	−2.7	−2.8	−3.1	−2.6
Crop protection—diseases	−3.5	−3.4	−2.4	−3.5	−3.4	−2.3
Integrated soil fertility management	−4.4	−7.8	−2.4	−4.3	−7.8	−1.8
Drought tolerance	−1.5	−0.4	−1.2	−1.5	−0.5	−1.3
Water harvesting	−0.2	−0.1	−0.5	−0.2	−0.1	−0.7
Drip irrigation	−0.7	−0.1	−0.2	−0.8	−0.1	−0.2
Sprinkler irrigation	−0.4	−0.1	−0.1	−0.4	−0.1	−0.1

Source: Authors.

Notes: No-till was not modeled for rice in DSSAT. Results reflect baseline rice values interacting with no-till for maize and wheat. A1B = greenhouse gas emissions scenario that assumes fast economic growth, a population that peaks mid-century, and the development of new and efficient technologies, along with a balanced use of energy sources; CSIRO = Commonwealth Scientific and Industrial Research Organisation's general circulation model; DSSAT = Decision Support System for Agrotechnology Transfer; MIROC = Model for Interdisciplinary Research on Climate.

result in the largest improvements for maize yield, NUE for rice, and no-till on wheat. The smallest effects across the crops are for the two irrigation technologies and water harvesting (Figure 5.1).

Table D.1 in Appendix D presents raw yields and yield growth rates for all regions.[2] For maize, the growth rate ranges from 0.29 percent to 3.9 percent per year. The lowest value is recorded for sprinkler irrigation and drip irrigation in North America, and the largest is for no-till in South Asia. In South Asia, eight technologies have high annual rates of growth of more than 2.4 percent (crop protection, drought tolerance, heat tolerance, ISFM, NUE, and no-till).[3]

For rice, annual growth yield rates range between 0.24 percent for advanced irrigation in the East Asia and Pacific region and 3.35 percent for

---

2 Yield growth rate is defined as a percentage (straight-line) growth rate; that is, it is a non-compound rate.

3 Refer to Chapter 3 (especially the section on assumptions and limits of the study) to learn about the limits regarding the implementation of crop protection.

FIGURE 5.1   Global yield impacts compared to the baseline scenario, by technology and crop, 2050 (%)

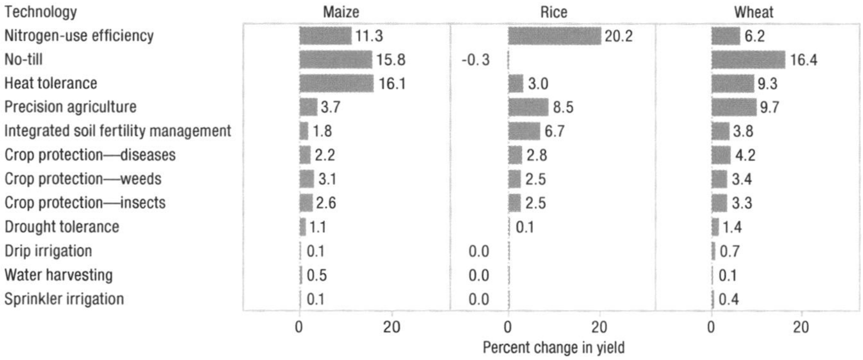

Technology	Maize	Rice	Wheat
Nitrogen-use efficiency	11.3	20.2	6.2
No-till	15.8	-0.3	16.4
Heat tolerance	16.1	3.0	9.3
Precision agriculture	3.7	8.5	9.7
Integrated soil fertility management	1.8	6.7	3.8
Crop protection—diseases	2.2	2.8	4.2
Crop protection—weeds	3.1	2.5	3.4
Crop protection—insects	2.6	2.5	3.3
Drought tolerance	1.1	0.1	1.4
Drip irrigation	0.1	0.0	0.7
Water harvesting	0.5	0.0	0.1
Sprinkler irrigation	0.1	0.0	0.4

Percent change in yield

Source: Authors.

Notes: No-till was not modeled for rice in DSSAT. Results reflect baseline rice values interacting with no-till for maize and wheat. DSSAT = Decision Support System for Agrotechnology Transfer.

NUE in SSA. For maize in South Asia, all technologies have annual growth rates near or greater than 2.4 percent. For wheat, the largest rate is 2.07 percent for no-till in SSA and the lowest is 0.04 percent for advanced irrigation in Western Europe.

The largest increases in yield compared to the baseline in 2050 are observed in South Asia under no-till for maize and wheat and under NUE for rice. The East Asia and Pacific region also shows substantial yield improvements, as does SSA under no-till for maize and wheat and under NUE for rice. We find a small negative yield effect under PA in SSA for maize, as higher potential for the technology in other regions reduces the comparative advantage for this technology in SSA (Table D.2 in Appendix D; Figure 5.2). The yield impacts are thus largely consistent with the biophysical modeling results presented in Chapter 4. Where changes are observed, they are due to differences in adoption ceilings as well as to interactions of the three cereal crops with other agricultural commodities, and resulting changes in food demand, supply, trade, and international food prices.

Changes in Cereal Production and Harvested Area under Alternative Technologies

Regardless of the technology adopted, most maize, rice, and wheat production (averaged across the three crops) largely comes from a small set of regions that

FIGURE 5.2   Yield impacts compared to the baseline scenario for selected regions, by technology and crop, 2050 (%)

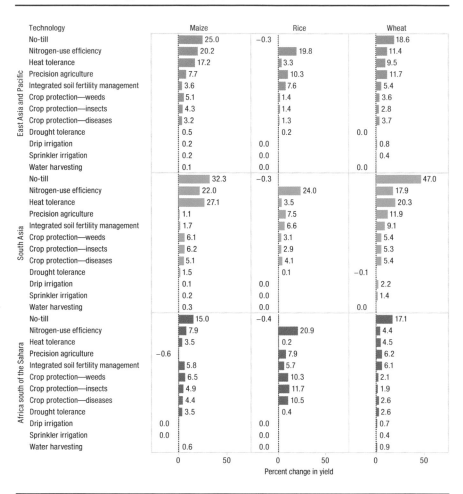

Source: Authors.

Notes: No-till was not modeled for rice in DSSAT. Results reflect baseline rice values interacting with no-till for maize and wheat. DSSAT = Decision Support System for Agrotechnology Transfer.

already provide the bulk of production under the baseline. The East Asia and Pacific region, North America, and South Asia are the three largest producing regions. Heat-tolerant varieties, NUE varieties, and no-till are the technologies that can shift production the most in these regions.

Globally, the largest production increases compared to the baseline in 2050 are achieved through no-till and heat tolerance for maize, NUE and PA

FIGURE 5.3   Global change in production compared to the baseline scenario, by technology and crop, 2050 (%)

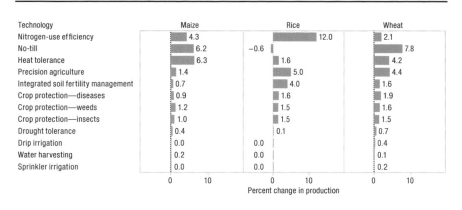

Source: Authors.
Notes: No-till was not modeled for rice in DSSAT. Results reflect baseline rice values interacting with no-till for maize and wheat. DSSAT = Decision Support System for Agrotechnology Transfer.

for rice, and no-till and PA for wheat (Figure 5.3). The largest increases in the production of maize compared to the baseline are in South Asia and in East Asia and Pacific for NUE and no-till, and in South Asia for heat-tolerant varieties. For rice, the largest positive effects are for NUE in South Asia and for wheat no-till, also in South Asia. Small production declines are seen for rice in North America, Europe, and central Asia (Table D.3 in Appendix D; Figure 5.4).

Technology-induced improvements in yields and production may result in increased land use intensity and thus less need to further expand harvested areas by 2050. As expected, the technologies with highest yield impacts also lead to the largest increase in supply and thus cause the largest savings in arable area expansion globally (Figure 5.5).

Heat-tolerant varieties and no-till lead to substantial decreases in harvested area in North America. Declines are also significant in Latin America and the Caribbean for the same technologies. For rice, the largest decrease in area results from the adoption of NUE in Western Europe, whereas no-till adoption leads to the largest decrease in harvested area of wheat in Latin America and the Caribbean. In fact, some of the largest declines in arable area are in Western Europe and Latin America and the Caribbean (Table D.4 in Appendix D; Figure 5.6).

FIGURE 5.4   Change in production for developing countries compared to the baseline scenario, by technology and crop, 2050 (%)

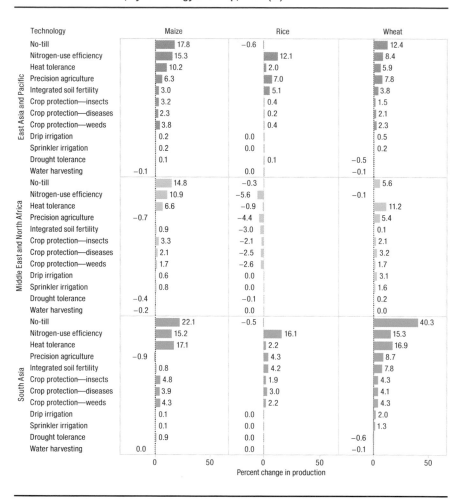

Source: Authors.
Notes: No-till was not modeled for rice in DSSAT. Results reflect baseline rice values interacting with no-till for maize and wheat. DSSAT = Decision Support System for Agrotechnology Transfer.

Changes in Net Cereal Trade under Alternative Technologies

Developing countries will continue their net import positions for wheat and maize between 2010 and 2050, as well as their net export position for rice (Table D.5 in Appendix D; Figure 5.7). Widespread adoption of no-till, NUE, and PA would considerably reduce net imports of wheat and maize compared to the baseline, whereas NUE would substantially boost rice exports. The

FIGURE 5.5 Global change in harvested area compared to the baseline scenario, by technology and crop, 2050 (%)

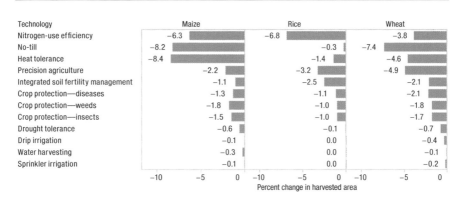

Source: Authors.
Notes: No-till was not modeled for rice in DSSAT. Results reflect baseline rice values interacting with no-till for maize and wheat. DSSAT = Decision Support System for Agrotechnology Transfer.

group of developed countries will remain a major exporter of maize and wheat, with net exports especially boosted by the adoption of heat-tolerant varieties.

We find that by 2050, SSA will still be a net importer of all three crops. Widespread expansion of ISFM in this region would be the best technology to reduce net imports of maize and wheat, along with advanced irrigation technologies. South Asia will be a net importer of wheat and maize but an exporter of rice under all technology adoption scenarios. Rice exports will be particularly boosted through large-scale adoption of NUE, whereas no-till will contribute the most to reducing imports of maize and wheat (Figure 5.8).

Changes in Food Security Outcomes under Alternative Technologies

The Food and Agriculture Organization of the United Nations estimates that between 1990 and 2013, the prevalence of undernourishment decreased by more than 6 percentage points globally and by more than 9 percentage points in developing countries and in the least developed countries.[4] Despite these improvements, 870 million people are still chronically undernourished today (von Grebmer et al. 2013). According to the 2013 Global Hunger Index

---

4 http://www.fao.org/economic/ess/ess-fs/fs-data/en/.

FIGURE 5.6   Change in harvested area compared to the baseline scenario for selected regions, by technology and crop, 2050 (%)

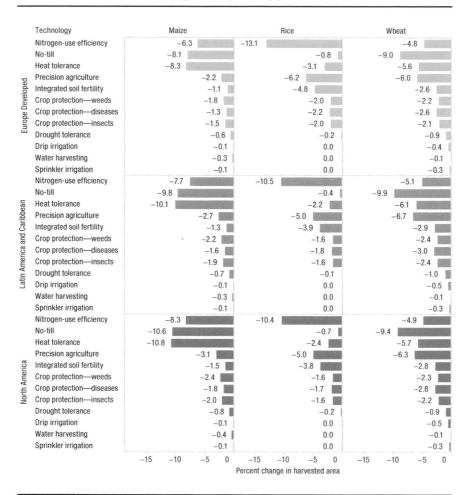

Source: Authors.

Notes: No-till was not modeled for rice in DSSAT. Results reflect baseline rice values interacting with no-till for maize and wheat. DSSAT = Decision Support System for Agrotechnology Transfer.

report, undernutrition decreased in all regions of the world, albeit at different rates, but conditions remain serious in SSA and alarming[5] in South Asia, overall the region with the worst Global Hunger Index score (von Grebmer et al. 2013).

---

5 The terms "serious" and "alarming" correspond to specific score ranges along the Global Hunger Index scale. For details, see von Grebmer et al. (2013).

FIGURE 5.7 Net trade of maize, rice, and wheat for developing countries, by technology, 2050 (thousand metric tons)

Technology
Maize
Rice
Wheat
Developing
No-till −30,237 1,029 −19,001
Drip irrigation −51,270 957 −30,156
Water harvesting −52,859 957 −31,639
Integrated soil fertility management −44,202 1,608 −29,223
Crop protection—diseases −43,929 1,749 −29,764
Precision agriculture −42,175 1,704 −27,477
Nitrogen-use efficiency −31,465 3,197 −23,811
Crop protection—insects −42,423 1,634 −31,340
Sprinkler irrigation −51,273 956 −30,828
Heat tolerance −75,064 1,373 −32,152
Crop protection—weeds −39,944 1,707 −29,067
Drought tolerance −50,315 992 −32,234
Baseline −51,472 955 −31,713
−100 0 100 −100 0 100 −100 0 100
Net trade (thousand metric tons)

Source: Authors.

Note: Negative numbers indicate net imports. Positive numbers indicate net exports.

FIGURE 5.8 Net trade of maize, rice, and wheat for selected regions, by technology, 2050 (thousand metric tons)

Technology
Maize
Rice
Wheat
South Asia
No-till −11,483 21,395 −24,749
Nitrogen-use efficiency −13,221 34,270 −42,156
Integrated soil fertility management −18,135 23,728 −51,234
Crop protection—weeds −16,752 23,417 −55,967
Crop protection—insects −16,375 22,992 −55,868
Crop protection—diseases −16,698 24,646 −56,860
Drip irrigation −18,109 21,515 −56,600
Sprinkler irrigation −18,089 21,508 −57,170
Baseline −18,112 21,502 −58,334
Water harvesting −18,287 21,490 −58,555
Heat tolerance −14,230 23,072 −46,066
Drought tolerance −18,008 21,411 −60,393
Precision agriculture −19,680 22,786 −55,424
Africa south of the Sahara
No-till −40,332 −29,325 −45,114
Nitrogen-use efficiency −40,288 −38,246 −42,215
Integrated soil fertility management −32,589 −32,973 −40,903
Crop protection—weeds −33,572 −28,608 −41,200
Crop protection—insects −34,236 −28,244 −41,186
Crop protection—diseases −34,170 −28,806 −41,540
Drip irrigation −35,248 −29,757 −40,142
Sprinkler irrigation −35,252 −29,753 −40,038
Baseline −35,146 −29,752 −39,884
Water harvesting −35,215 −29,734 −39,822
Heat tolerance −48,235 −31,470 −43,427
Drought tolerance −33,719 −29,742 −40,197
Precision agriculture −38,969 −33,751 −43,744
−100 −50 0 50 −100 −50 0 50 −100 −50 0 50
Net trade (thousand metric tons)

Source: Authors.

Note: Negative numbers indicate net imports. Positive numbers indicate net exports

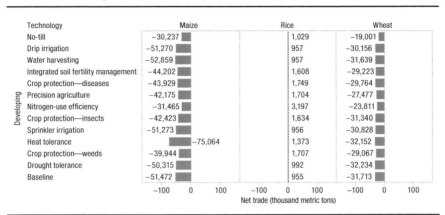

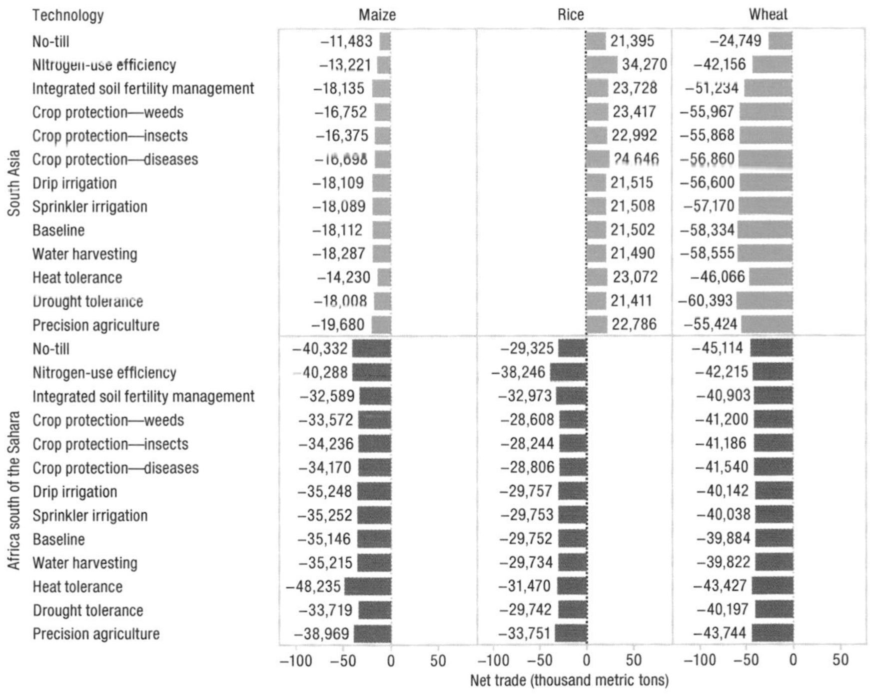

TABLE 5.5   Change in per capita kilocalorie availability compared to the baseline scenario, by technology, 2050 (%)

Technology	Developing countries	South Asia	Africa south of the Sahara	World
Nitrogen-use efficiency	2.62	2.62	3.19	2.38
No-till	1.98	2.12	2.29	1.93
Precision agriculture	1.73	1.94	1.72	1.63
Heat tolerance	1.64	1.59	2.17	1.55
Integrated soil fertility management	0.97	1.06	1.01	0.90
Crop protection—diseases	0.68	0.77	0.69	0.65
Crop protection—weeds	0.62	0.67	0.68	0.58
Crop protection—insects	0.59	0.65	0.64	0.56
Drought tolerance	0.18	0.20	0.19	0.18
Drip irrigation	0.07	0.09	0.06	0.07
Sprinkler irrigation	0.04	0.05	0.04	0.04
Water harvesting	0.03	0.02	0.05	0.03

Source: Authors.

As highlighted above, most of the technology adoption scenarios show increases in production and yields, and although all induce lower food prices in 2050 compared to the baseline, the largest declines are caused by NUE, heat tolerance, PA, and no-till. As a result, these technologies are also those that have the largest positive effects on calorie availability (Table 5.5), child malnutrition levels (Figure 5.9), and the population at risk of hunger (Figure 5.10), compared to the baseline.

In terms of improvements for calorie availability, SSA benefits most from the agricultural technologies evaluated; heat-tolerant varieties, NUE, and no-till have particularly large effects, with increases of 2–3 percent compared to the baseline. NUE provides a nearly 3 percent increase in calorie availability in South Asia and a more than 3 percent improvement in East Asia and the Pacific (Table D.6 in Appendix D; Figure 5.11).

In terms of percentage reduction in the number of malnourished children (between the ages of 0 and 5) compared to the baseline, the largest reductions are in Latin America and the Caribbean and in the Middle East and North Africa (Table D.7 in Appendix D; Figure 5.12). However, in terms of raw numbers, the largest reductions are in South Asia and especially in SSA, where the adoption of heat-tolerant varieties, NUE,

FIGURE 5.9   Change in the number of malnourished children in developing countries
compared to the baseline scenario, by technology, 2050 (%)

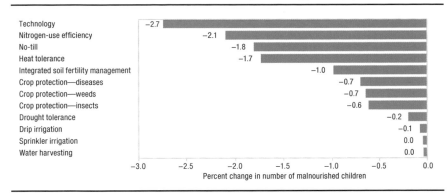

Source: Authors.
Note: Numbers are for children malnourished between the ages of 0 and 5.

FIGURE 5.10   Change in number of people at risk of hunger in developing countries
compared to the baseline scenario for selected regions, by technology,
2050 (%)

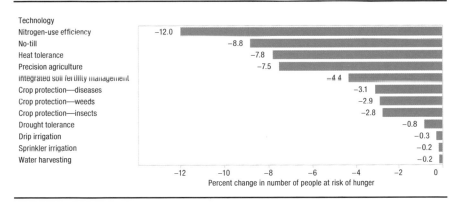

Source: Authors.

and no-till could reduce the number of malnourished children by more than
1 million compared to the baseline (for example, 43 million under no-till
versus 44 million under the baseline).

Overall, NUE and no-till are the technologies with the strongest reductions in the total population at risk of hunger across developing countries
(Table D.8 in Appendix D; Figure 5.12).

FIGURE 5.11   Change in kilocalorie availability per person per day compared to the baseline scenario for selected regions, by technology, 2050 (%)

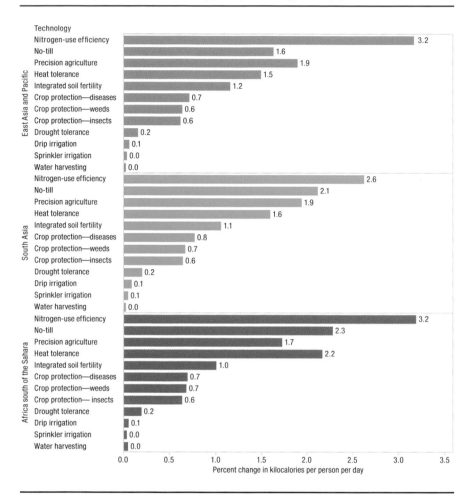

Source: Authors.

Results for Stacked Technologies—Simultaneous Adoption of Multiple Technologies: MIROC A1B Climate Change Scenario

We compare the impacts of the various agricultural technologies selected for this study with reference to a baseline to isolate the impacts of each technology and for unbiased comparisons. In farmers' fields, however, multiple technologies might well be adopted on a single piece of land or portions thereof. It is also realistic to expect that when these technologies

FIGURE 5.12   Change in the number of malnourished children compared to the baseline scenario for selected regions, by technology, 2050 (%)

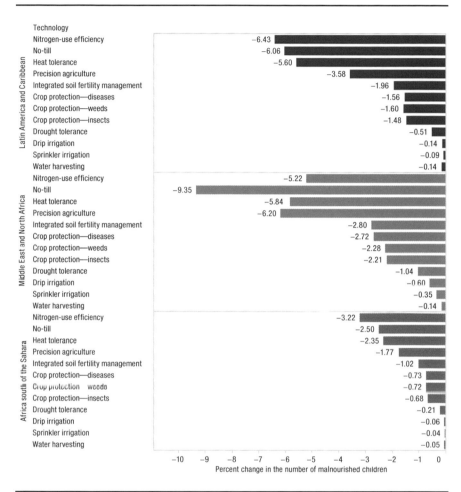

Source: Authors.
Note: Numbers are for children malnourished between the ages of 0 and 5.

become available and widely disseminated to farmers, each farmer would have the option to select a single technology or a combination of them that he or she sees fit. Farmers can even choose to adopt all technologies on parts of their farms.

Only a few of the technologies assessed could technically not be adopted simultaneously, for example, drip and sprinkler irrigation, but both can still

TABLE 5.6 Effects of stacked technologies on world prices of maize, rice, and wheat, compared to the baseline scenario, 2050 (%)

Technology stack[a]	Maize		Rice		Wheat	
	Price difference	Incremental contribution of technology	Price difference	Incremental contribution of technology	Price difference	Incremental contribution of technology
No-till	−15.8	−15.8	—	—	−15.5	−15.5
Drought tolerance	−16.7	−1.0	−3.6	−3.6	−16.6	−1.2
Heat tolerance	−30.1	−13.4	−9.3	−5.7	−25.2	−8.6
Nitrogen-use efficiency	−40.8	−10.7	−28.0	−18.7	−32.5	−7.3
Integrated soil fertility management	−42.8	−2.0	−33.8	−5.8	−36.1	−3.5
Precision agriculture	−46.2	−3.4	−41.0	−7.2	−42.4	−6.3
Water harvesting	−46.4	−0.2	—	—	−42.5	−0.1
Sprinkler irrigation	−46.5	−0.1	—	—	−42.8	−0.3
Drip irrigation	−46.7	−0.1	—	—	−43.4	−0.6
Crop protection— insects[b]	−48.3	−1.6	−43.0	−1.9	−44.8	−1.4
Crop protection— diseases[b]	−48.0	−1.3	−43.2	−2.2	−45.3	−1.9
Crop protection— weeds[b]	−48.6	−1.9	−43.0	−2.0	−44.9	−1.5

Source: Authors.

Note: — = not applicable.

[a]Each technology is stacked over (added to) the previous combination from top to bottom. Thus, drought tolerance is added to no-till; heat tolerance is added to the stack of no-till and drought tolerance, and so on.

[b]Each crop protection technology is added to the technology stack of drip Irrigation and the technologies before it for maize and wheat, and to the precision agriculture stack for rice.

be used in separate sections of the farm or in different cropping seasons. The same could be said of crop protection for diseases, arthropods, and weeds. It is not advisable to adopt all three when an area is only susceptible to one type of infestation.

We therefore anticipate that these technologies would be adopted in various combinations globally. This section explores the results of simulating such a multiple-adoption scenario for several or all the technologies at the global scale and their combined impacts on prices and food security. Combining them, one on top of the other (that is, stacking them) in the order of crop production schedules (that is, first land preparation, planting, and crop/farm management, followed by irrigation, and so forth) also makes it possible to

FIGURE 5.13   Price effects of stacked technologies compared to the baseline scenario, by crop and technology, 2050 (%)

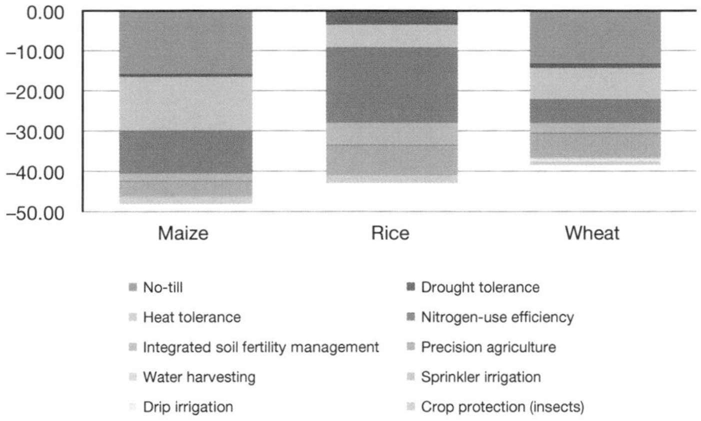

Source: Authors.

Note: A1B = greenhouse gas emissions scenario that assumes fast economic growth, a population that peaks mid-century, and the development of new and efficient technologies, along with a balanced use of energy sources; MIROC = Model for Interdisciplinary Research on Climate.

show the *marginal* contribution of each technology to the overall impacts of the stacked technologies.

Prices of cereals and also meat would be directly affected by improvements in cereal productivity. In this section, we only report prices of cereals (maize, rice, and wheat) that are directly affected by production improvements caused by the technologies.

Table 5.6 presents the effect of stacked technology adoption on world prices of cereals. In a global scenario where multiple combinations of the technologies are implemented on farmers' fields, production can be high enough to cut world prices of maize by up to 49 percent, up to 43 percent for rice, and 45 percent for wheat. Incremental contributions are highest for heat-tolerant varieties of maize, NUE for rice, and no-till for wheat—which are, not surprisingly, also the technologies with large impacts on yields (Figure 5.13).

Crop protection technologies, though showing moderate contributions individually, when taken together can be major contributors to reducing cereal prices—by as much as 5 percent for wheat and maize and 6 percent for rice.

The impacts of these stacked technologies on global food security are presented in Table 5.7. The number of malnourished children and of people at risk of hunger are the indicators used to measure improvements that may

TABLE 5.7   Effects of stacked technologies on global food security compared to the
baseline scenario, 2050

Technology stack[a]	Malnutrition			Food insecurity		
	Number of malnourished children in 2050 (millions)	Change from baseline (%)	Incremental contribution of technology (%)	Number of people at risk of hunger (millions)	Change from baseline (%)	Incremental contribution of technology (%)
Baseline	116.77	—	—	1,087.48	—	—
No-till	114.46	−1.97	−1.97	999.64	−8.08	−8.08
Drought tolerance	114.23	−2.17	−0.20	991.26	−8.85	−0.77
Heat tolerance	112.10	−4.00	−1.83	913.12	−16.03	−7.19
Nitrogen-use efficiency	108.82	−6.81	−2.81	800.18	−26.42	−10.39
Integrated soil fertility management	107.63	−7.83	−1.02	766.88	−29.48	−3.06
Precision agriculture	105.48	−9.67	−1.84	711.68	−34.56	−5.08
Water harvesting	105.26	−9.86	−0.19	710.16	−34.70	−0.14
Sprinkler irrigation	105.11	−9.98	−0.12	709.20	−34.78	−0.09
Drip irrigation	104.97	−10.10	−0.12	708.23	−34.87	−0.09
Crop protection—insects[b]	104.75	−10.29	−0.19	690.62	−36.49	−1.62
Crop protection—diseases[b]	104.57	−10.44	−0.34	688.08	−36.73	−1.85
Crop protection—weeds[b]	104.67	−10.36	−0.26	689.37	−36.61	−1.73

Source: Authors.

Note: — = not applicable; A1B = greenhouse gas emissions scenario that assumes fast economic growth, a population that peaks mid-century, and the development of new and efficient technologies, along with a balanced use of energy sources; MIROC = Model for Interdisciplinary Research on Climate.

[a]Each technology is stacked over (added to) the previous combination from top to bottom. Thus, drought tolerance is added to no-till; heat tolerance is added to the stack of no-till and drought tolerance, and so on.

[b]Each crop protection technology is added to the technology stack of drip Irrigation and the technologies before it for maize and wheat, and to the precision agriculture stack for rice.

result from simultaneous adoption of combinations of these technologies. With the use of these technologies, the number of malnourished children can be reduced by 12 percentage points—from a baseline projection to 2050 of 117 million children to 103 million children. The highest single contributor to the decline is NUE (3 percent), followed by no-till (2 percent).

The number of people at risk of hunger could be reduced by as much as 40 percent with the use of combinations of these technologies—from a baseline projection of more than 1 billion people to 0.7 billion. Again, NUE and other technologies with large effects on yields are the greatest contributors.

This stacked-technology simulation suggests that opportunities exist for complementarity of these technologies. Where combinations of technologies are adopted on farms, production and profitability for producers can be increased, prices for consumers can be lowered, and nutrition and food security of the population can be substantially improved.

Results for Alternative Technologies: Comparison of MIROC A1B and CSIRO A1B Climate Change Scenarios

The CSIRO A1B and MIROC A1B climate change scenarios are at opposite ends of the spectrum of projected changes in temperature and precipitation. CSIRO shows a small (0.7 percent) increase in precipitation by 2050 compared to an overall increase of 4.7 percent for MIROC. At the same time, the temperature increase in MIROC is higher (2.8°–3.0°C compared to 1.4°–1.6°C for CSIRO) (Nelson et al. 2010).

Despite these differences, the effects of alternative agricultural technologies across the climates are remarkably similar. The small overall differences are evident in a graph comparing changes in calorie availability for developing countries (Figure 5.14). The largest difference exists for heat-tolerant varieties,

FIGURE 5.14 Change in kilocalorie availability per person per day compared to the baseline scenario for developing countries, by technology, MIROC A1B and CSIRO A1B scenarios, 2050 (%)

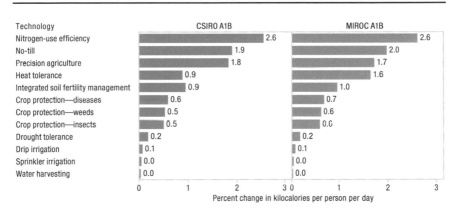

Source: Authors.
Note: A1B = greenhouse gas emissions scenario that assumes fast economic growth, a population that peaks mid-century, and the development of new and efficient technologies, along with a balanced use of energy sources; CSIRO = Commonwealth Scientific and Industrial Research Organisation's general circulation model; MIROC = Model for Interdisciplinary Research on Climate.

with larger calorie availability under MIROC A1B compared to the drier and cooler CSIRO scenario. Yield changes are also similar for the two climate change scenarios, except for heat-tolerant varieties, where variations are larger (Figure 5.15).

FIGURE 5.15   Change in yield compared to the baseline scenario for developing countries, by technology, MIROC A1B and CSIRO A1B scenarios, 2050 (%)

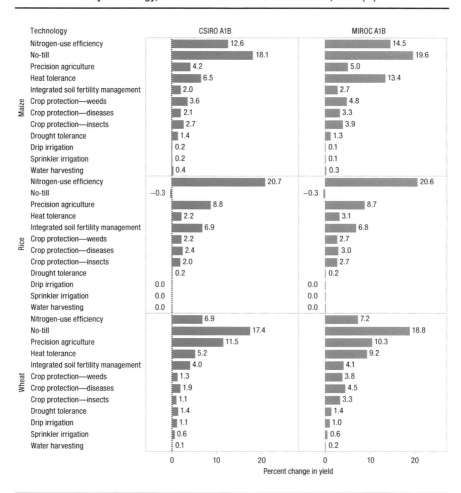

Source: Authors.
Notes: No-till was not modeled for rice in DSSAT. Results reflect baseline rice values interacting with no-till for maize and rice. A1B = greenhouse gas emissions scenario that assumes fast economic growth, a population that peaks mid-century, and the development of new and efficient technologies, along with a balanced use of energy sources; CSIRO = Commonwealth Scientific and Industrial Research Organisation's general circulation model; DSSAT = Decision Support System for Agrotechnology Transfer; MIROC = Model for Interdisciplinary Research on Climate.

Chapter 6

Implications for Technology Investment

I n 2011, the world's population reached 7 billion. Over the next four decades, an extra 2 billion people will be added, nearly all in low- and medium-income developing countries. These countries already face serious challenges in satisfying basic needs, including the provision of food, water, and energy. As a result of continued population and income growth, we project that nearly 80 percent more meat, 52 percent more cereals, and 40 percent more roots and tubers will be needed between 2005 and 2050 under the MIROC A1B scenario used in this study, with adverse consequences for the world's poor and vulnerable populations. Under the same baseline scenario, prices for maize, rice, and wheat would increase by 104 percent, 79 percent, and 88 percent, respectively, and the number of people at risk of hunger in the developing world would grow from 881 million in 2005 to 1,031 million people by 2050 (IFPRI IMPACT baseline, MIROC A1B used in this analysis). Climate change is a significant contributor to the projected higher prices and could decrease maize yields by 9–18 percent in 2050 compared to a no–climate change scenario, depending on climate change scenario, on cropping system (rainfed or irrigated), and on whether the carbon fertilization effect is included; rice yields could drop by 7–27 percent; and wheat yields are projected to decline by 18–36 percent (Nelson et al. 2009).

At the same time, land scarcity is rapidly worsening, and land degradation continues apace. Water scarcity and degradation are also increasing due to economic and population growth, poor water management, and the impacts of climate change (MA and WRI 2005; Anseeuw et al. 2012).

Addressing the challenges of climate change; growing land, water, and energy scarcity; rising long-term food prices; and poor progress in improving food security will require increased food production without further damage to the environment. Accelerated investments in agricultural R&D will be crucial to support food production growth. However, the specific set of agricultural technologies that should be brought to bear remains highly uncertain. The future technology mix will have major impacts on agricultural production, food consumption, food security, trade, and environmental quality in developing countries.

This study addresses these issues based on detailed analyses of a series of agricultural technologies that have the potential for broad application and (further) substantial expansion. The selected technologies, identified based on in-depth literature reviews and consultations with a large number of scientists and experts, range from traditional management practices to modern farm applications and new crop varieties and include OA; no-till; ISFM; water harvesting; drip irrigation; sprinkler irrigation; PA; heat tolerance; drought tolerance; NUE; and crop protection from diseases, insects, and weeds. Several of these technologies have already been partially adopted in some parts of the world, such as no-till and drip irrigation. Others are in the final stages of development and field trials. All can be rolled out in one form or another across large agricultural areas if appropriate investments, support policies, and institutions (all of which have associated costs) are put in place. None of these technologies are blue-sky thinking. Are these the only technologies that will matter over the next 40 years? This is unlikely, but we believe that the broad range of technologies illustrates the key strands of R&D that matter in the face of growing natural resource scarcity, climate change, and higher food demands.

We compare the global and regional effects of these different technologies on yields and resource use (that is, nitrogen leaching and water savings) for three major cereal crops: rice, maize, and wheat. The study uses a groundbreaking crop modeling approach (through the DSSAT model) that provides comprehensive data across a 60-kilometer by 60-kilometer grid of global arable land. This unprecedented level of detail demonstrates how each technology would impact yields of maize, rice, and wheat at a granular level over 40 years under two different climate change scenarios to 2050, when climate change impacts on cereals are likely to be substantial and food demands much higher than they are today. The crop modeling results are then input into IFPRI's IMPACT model, assuming adoption pathways that reflect to some extent perceptions of profitability, initial costs and capital, risk-reduction, and complexity of the technology, to simulate global food supply and demand, food trade, and international food prices for these three crops, as well as the resulting number of people at risk of food insecurity.

Any complex and multidimensional modeling effort, such as the present one, is obliged to make various simplifying assumptions, especially when attempting to estimate the future development of plausible scenarios (details provided in Chapter 3). Consequently, simulated outcomes are not intended to be taken at face value but rather to demonstrate possible orders of magnitude that suggest areas to be studied further as more data and yet more advanced analytical models are developed.

The study is geared toward policymakers in ministries of agriculture, national agricultural research institutes, multilateral development banks, the private sector, and CGIAR. In particular, policymakers are looking for guidance on which technology strategies they should pursue to secure national, regional, and global food security in a world of growing natural resource scarcity and competition for land, water, and energy across productive sectors and even increasingly across borders.

Model Results

Based on the biophysical (DSSAT) model results, under the hotter, wetter MIROC A1B climate change scenario, the largest ex ante yield impacts for maize are achieved with heat tolerance, followed by no-till. NUE has the highest yield impact for rice, followed by ISFM. For wheat, no-till has the highest yield impact, followed by PA. In contrast, under the less hot and drier CSIRO A1B climate scenario, the benefits of heat tolerance are lower, moving this technology into third place for maize globally. The combined impact across three types of crop protection (insects, pests, and weeds) ranks fourth for each crop, although the simulation of these technologies is still based on a rough approximation.

When adoption profiles that specify adoption ceilings by agricultural technology, trade in agricultural commodities, changes in international food prices, and other economic relationships are taken into account in IMPACT, the projected global improvements in yields in 2050 for maize are 16 percent for heat-tolerant varieties and no-till; for rice, 20 percent for NUE and 9 percent for PA; and for wheat, 16 percent for no-till, 10 percent for PA, and 9 percent for heat-tolerant varieties. Combining crop protection for diseases, arthropods, and weeds enhances yields of maize by 8 percent, rice by 8 percent, and wheat by 11 percent.

We also find a particularly large range of technologies with high potential for the breadbaskets of South Asia.

The effects on food security of the technologies studied (again using the MIROC A1B scenario) could be substantial. The technologies evaluated in this study increase average calorie availability in Africa south of the Sahara the most; heat-tolerant varieties, NUE, and no-till have particularly large effects for calorie availability, with increases of 2–3 percent compared to the baseline. NUE provides nearly 3 percent more available calories in South Asia and more than 3 percent in East Asia and the Pacific.

The number of food-insecure people in developing countries in 2050 could be reduced by 12 percent if NUE technologies were successfully developed and adopted, by 9 percent if no-till were adopted more widely, and by 8 percent with widespread adoption of heat tolerance and PA. Adoption of the three types of crop protection together is projected to reduce the number of food-insecure people by close to 9 percent, and ISFM would reduce their number by 4 percent. Both NUE and no-till address important soil quality constraints that are particularly important in developing countries. In terms of making inroads on child undernutrition, widespread adoption of heat tolerance, NUE, and no-till each could reduce the number of malnourished children by more than 1 million compared to the baseline.

Other important findings are summarized as follows. For several technologies across the three crops, the largest relative yield gains are in SSA, South Asia, and parts of Latin America and the Caribbean. First, agricultural technology impacts differ substantially by region and within regions by country. Given the heterogeneity in yield response, it is therefore important to target specific technologies to specific regions and countries. Subject to the assumptions of the simulations presented in this book, such targeting includes heat tolerance in North America and South Asia; drought tolerance in Latin America and the Caribbean, the Middle East and North Africa, and SSA; and crop protection in Eastern Europe, South Asia, and SSA. PA shows highest total gains in major production areas in the Middle East and North Africa, South Asia, and parts of Western Europe and has strongest yield impacts for wheat. NUE is critical to reduce resource use for sustainable development and improves yields substantially in most developing regions, particularly in South Asia, East Asia and the Pacific, and SSA. The largest potential for ISFM is in low-input regions in Africa, in South Asia, and in parts of East Asia and the Pacific.

Second, the effects of agricultural technology are amplified with irrigation. Although direct yield impacts from substituting furrow irrigation with drip and sprinkler irrigation are small for maize and wheat, water savings are substantial, indicating that yield levels can be sustained in a given area while releasing water for use elsewhere. Moreover, as yield impacts of other technologies tend to be larger with irrigation, continued investment in cost-effective irrigation should go hand in hand with technology rollout.

Third, technologies are important for addressing abiotic stresses that are expected to increase as a result of climate change. Drought-tolerant varieties perform as well as susceptible varieties under no drought stress and have significant yield benefits under drought conditions. Heat-tolerant varieties can also help reduce the projected adverse effects of climate change. In addition to biotic stresses, successful development of cost-effective crop protection from weeds, insects, and diseases may reap large benefits in most developing countries, whereas the scope for additional benefits from adoption is limited in the developed world because of the high levels of crop protection already achieved.

Recent global studies have shown that, on average, OA achieves lower yields compared to conventional systems. Although there is still some uncertainty as to the magnitude of this difference from crop to crop, the consensus is that yields for cereals are significantly lower. As a result, expansion of OA would require considerably more land to achieve production increases. As highlighted by Seufert, Ramankutty, and Foley (2012, 229), "although yields may be only part of a range of ecological, social, and economic benefits delivered by farming systems, it is widely accepted that high yields are central to sustainable food security on a finite land basis." The lower yields and the additional land requirements necessary to both increase production and maintain fertility in fields using OA put in question the scalability of this option and its value as a solution for food production at the global level. Moreover, although environmental externalities are lower for OA per unit area, they are significantly larger when measured per unit of output (Balmford, Green, and Phalan 2012; Tuomisto et al. 2012). Despite this, some benefits from OA can be and are replicated in conventional agriculture through the use of knowledge-intensive management processes, such as PA and minimum or no-till.

The results suggest that sustainably meeting the challenge of climate change while substantially improving food security requires a three-pronged effort: increased crop productivity through enhanced investment in agricultural research, development and use of resource-conserving management, and increased investment in irrigation. Crop breeding should target abiotic stresses (such as heat and drought) and biotic stresses (such as pests and diseases), as well as continuing to invest in broad-based yield improvement. Research focused on resource-conserving management and technology should be expanded, including no-till and minimum tillage, ISFM, improved crop protection, and PA. Increased investment in cost-effective irrigation will serve to increase the returns to other technologies, and such advanced irrigation technologies as drip and sprinkler irrigation can save water in specific locations while maintaining yield levels.

Policies and Institutions for Technology Implementation

Much has been written on the policies, institutions, and investments conducive to accelerated discovery, development, and diffusion of agricultural technologies. This book does not summarize the literature, but instead provides a short overview of key areas that need to be considered when moving toward accelerated technology development and uptake for sustainable agriculture and food security. Policy advances will be required in all areas of the technology development cycle and need to be combined with appropriate institutions and governance mechanisms to continually improve the final technology outcomes.

The most common policy measures aimed at promoting technology adoption directly are

1. investments in agricultural R&D and extension services;

2. provision of incentives to private developers of technology when the appropriation of rewards is more difficult, or to early adopters, or as a second-best policy when other issues that constrain adoption (for example, the absence of sustainable finance services) cannot be overcome in the short and medium term; and

3. expansion of sustainable finance arrangements to farmers for investing in technologies.

For drip and sprinkler irrigation, promoting farmers' associations, improving market access, and sustaining supportive financial institutions are important policies. Biosafety laws and regulations play a dominant role in the adoption of genetically modified crops. Some policy measures can have an indirect—positive or negative—influence on technology adoption. This is the case for the provision of subsidies for agricultural inputs or machinery not directly related to a specific technology. For instance, subsidies for herbicides can increase adoption of no-till, because no-till requires more weed control than conventional tillage, but potentially adverse effects on the environment need to be addressed. Subsidies to early adopters can also promote the use of inputs, such as fertilizer and improved seeds, but many of the benefits of subsidies tend to be captured by richer farmers who have the highest effective demand for inputs. Subsidies can also result in overuse of inputs and can distort markets and inhibit efforts to develop effective seed and fertilizer markets and distribution systems. Land use policies are also an example of such indirect impacts: in China and India, the ban on burning of residues has resulted in the increased spread of minimum tillage (Singh et al. 2010; Wang et al. 2010).

Although public investment in agricultural research has historically driven technological change in developing-country agriculture, recent trends suggest that the private sector will need to play a larger role in the future

(Pray and Fuglie 2001; Pray, Fuglie, and Johnson 2007; Fuglie et al. 2011). Expansion of private investment in research and dissemination of new varieties will be especially crucial for developing new varieties with traits for heat tolerance, drought tolerance, and NUE. Although there is optimism about the private sector's ability to generate new technologies relevant to small-holders in developing-country agriculture, current levels of private investment remain low because of constraints relating to policies, incentives, and markets (Naseem, Spielman, and Omamo 2010; Beintema et al. 2012). For development of these advanced varietal technologies, improved public regulatory frameworks and strengthened institutions to provide appropriate incentives for plant breeding, product development, and diffusion will be essential. Although quantification of the effects of policies on technology adoption is rare in the literature, it is clear that these effects vary dramatically. For instance, subsidizing sprinklers in China by 45 percent of total sprinkler cost led to a 10 percent increase in the adoption rate of sprinkler irrigation (Yu and Jensen 2010), whereas the distribution of conventionally improved seeds in India was reported to have a negligible impact on the adoption of these seeds by farmers (Mackill et al. 2010).

Some incentives can even negatively affect technology adoption, owing to the disincentives they introduce with respect to other worthy technologies. Examples include the provision of input subsidies for fertilizer, water, and rural electricity for irrigation (Chadha and Davenport 2011), which reduces the potential benefits of NUE and water-saving technologies, and also diminishes the likelihood of adoption of no-till to save on the costs of these subsidized inputs. Thus, for policy measures to effectively promote the adoption of new technologies, full consideration of the direct and indirect impacts is necessary.

Another key condition to ensure effectiveness of policies is that they be targeted to the needs, preferences, resources, and environments of farmers. An example of a failure to do so is described in a review by Belder et al. (2007) of the distribution of low-head, low-cost drip irrigation kits to vulnerable households facing acute hunger and suffering from HIV/AIDS in Zimbabwe after 2002. The review of drip irrigation kit adoption and outcomes in 14 districts in Zimbabwe showed that after 2 years, only about one-third of the kits were still in operation; after 3 years, the number had dropped to one-sixth of the original number of kits handed out. In addition, the drip kits did not achieve their intended purposes. They did not save labor or water, as farmers tended to continue to apply bucket irrigation in parallel; they also did not improve nutritional outcomes, as farmers with bucket irrigation had more diverse gardens. The study concludes that vulnerable people were a poor target of the technology, because it increased labor (particularly high water lift) and was risky and relatively complex.

To evaluate whether specific policies are achieving their objective of boosting technology dissemination, and to be able to arrange for corrective adaptations when necessary, monitoring systems are needed. The cost of such systems will be offset by the savings generated by eliminating expenditures on policies that prove ineffective.

Given that many of the technologies are highly knowledge intensive, extension systems must increase knowledge capacity, and innovative forms of extension—through information and communication technologies, for example—should be implemented. This is particularly important for technology combinations (such as no-till with PA) that we identified as having substantial yield benefits in most regions.

Another aspect that needs to be considered when supporting technology advancement is that some of the technologies discussed are scale dependent. For such technologies to take hold, farmer-led institutions that can jointly invest in these technologies must increase in number and spread.

Moreover, several technologies will take many years to reap final benefits. This time lag often hinders adoption in places where land tenure systems are weak or farmers do not have access to cheap financing. Such technologies include no-till, ISFM, and water harvesting, which have been shown to yield most benefits after being used for several years.

Summary and Limitations

This book helps fill an important information gap on the yield potential of a range of agricultural technologies over a 40-year span. It considers highly disaggregated biophysical information as well as food supply, demand, and price and trade relationships under growing natural resource scarcity and climate change. The book assesses future scenarios of the potential impact and benefits of agricultural technologies in terms of future yield and production growth, food security, demand, and trade. Comprehensive policy and impact scenario analysis can contribute to the understanding of the role of alternative technologies in an integrated fashion to support specific agricultural sector policies and investment strategies.

Although this study has made important advances in ex ante technology assessments with a focus on highly disaggregate characterization of a wide range of agricultural technologies and linkage to agricultural sector models at global scale, additional research in several areas could further improve the results. First, characterization of technologies and parameterization of this characterization in crop models, a decisive step in the assessment, could benefit from yet further analysis and even wider consultations. Even though all technologies were characterized to reflect their reality on the ground as feasible (or as envisioned in their respective crop development stages), DSSAT limitations prevented process-based modeling of OA and crop protection technologies. We therefore relied on published literature for yield impacts and disaggregated these impacts as feasible by region and under future climates as described in Chapter 3. Moreover, because organic maize, rice, and wheat crops have much lower yield potential than conventional high-yield varieties, we did not consider OA in IMPACT. The yield increase test was the first step in the analysis that organically grown cereals did not pass. Moreover, for this study, we consulted with expert agronomists from the public and private sectors on technology specifications. We also worked with the IFPRI-led CGIAR Global Futures initiative on the characterization of drought tolerance, wheat cultivars, and wheat locations. We also received feedback from more than 400 agriculture experts on some technology characteristics of maize, rice, and wheat. Finally, we held expert meetings in Brazil (for Latin America) and India (for South Asia) as well as with several life sciences companies on technology characterization and preliminary results.

Uncertainties are particularly large for crop protection technologies, because estimates are based on climatic favorability for a limited set of representative species. Thus, the crop modeling yield results in this study that include crop protection data represent scenarios of yield changes based on simple models of pests and their responses to climate and pest management in the form of pesticide use. Our use of pesticide data is meant to illustrate the importance of crop protection for yields as part of any good management system. It is not our intent to evaluate or recommend a policy of worldwide adoption of pesticides. We did not have the opportunity to evaluate alternative crop protection strategies or combinations of crop protection methods in our analyses. For many parts of the world, the study lacked country- and region-specific data on crop losses to pests and the effects of alternative crop protection strategies. It is our hope that this work serves as a stepping stone and opens the door to more policy discussion of crop protection in the context of climate change, crop production, and global food security in the coming decades.

Second, the reliability of the study's results would benefit from including cost estimates related to technology discovery, development, and dissemination in the field. This study used coarse adoption profiles based on the expertise of selected study authors to reflect cost concerns, among other socioeconomic factors. Use of adoption profiles remains a second-best approach to in-depth analysis of the cost of technology. Thus, a separate, in-depth analysis of cost estimates by region would be needed to complement the analysis. Similarly, we did not assess what other expenditures (private or public) will be involved in bringing about the changes in technology that are envisioned in the scenarios analyzed. Specific elements include expenditures on agricultural R&D and extension, or the provision of infrastructure that changes farmer costs of adopting particular technologies. Such work is a research project in its own right and needs to be advanced in future studies.

Third, this analysis is focused on yield potential and food security outcomes under long-term climate change. A more complete study would also include scenarios under increased climate extremes, which are not fully reflected in currently available climate scenarios. This analysis approximated climate extremes (drought) through a special analysis of drought-tolerant technology, but more work on other technologies is needed.

Fourth, although this analysis is focused on environmental impacts of nitrogen emissions and water savings, more work needs to be done to assess the greenhouse gas emissions and the energy requirements of these technologies.

References

This reference list contains references for this book and for its appendixes, which are available online at http://www.ifpri.org/publication/food-security-world-natural-resource-scarcity.

Aggarwal, P. K., S. K. Bandyopadhyay, H. Pathak, N. Kalra, S. Chander, and S. Kumar. 2000. "Analysis of Yield Trends of the Rice-Wheat System in North-Western India." *Outlook on Agriculture* 29 (4): 259–268.

Ainsworth, E. A., and D. R. Ort. 2010. "How Do We Improve Crop Production in a Warming World?" *Plant Physiology* 154 (2): 526–530.

Alam, M., T. P. Trooien, T. J. Dumler, and T. H. Rogers. 2002. "Using Subsurface Drip Irrigation for Alfalfa." *Journal of the American Water Resources Association* 38 (6): 1715–1721.

Alene, A. D., A. Menkir, S. O. Ajala, B. Badu-Apraku, A. S. Olanrewaju, V. M. Manyong, and A. Ndiaye. 2009. "The Economic and Poverty Impacts of Maize Research in West and Central Africa." *Agricultural Economics* 40: 535–550.

Alston, J. M., J. M. Beddow, and P. G. Pardey. 2009. "Agricultural Research, Productivity, and Food Prices in the Long Run." *Science* 325 (5945): 1209–1210.

Anseeuw, W., M. Boche, T. Breu, M. Giger, J. Lay, P. Messerli, and K. Nolte. 2012. *Transnational Land Deals for Agriculture in the Global South: Analytical Report Based on the Land Matrix Database.* Bern, Switzerland: Centre for Development and Environment; Montpellier, France: CIRAD; and Hamburg, Germany: German Institute of Global and Area Studies.

Ashraf, M. 2010. "Inducing Drought Tolerance in Plants: Recent Advances." *Biotechnology Advances* 28 (1): 169–183.

Badgley, C., J. Moghtader, E. Quintero, E. Zakem, M. J. Chappell, K. Aviles-Vazquez, A. Samulon, and I. Perfecto. 2007. "Organic Agriculture and the Global Food Supply." *Renewable Agriculture and Food Systems* 22 (2): 86–108.

Badgley, C., and I. Perfecto. 2007. "Can Organic Agriculture Feed the World?" *Renewable Agriculture and Food Systems* 22 (2): 80–82.

Balmford, A., R. Green, and B. Phalan. 2012. "What Conservationists Need to Know about Farming." *Proceedings of the Royal Society* B 279: 1–12. Available at http://rspb.royalsocietypublishing.org/content/early/2012/04/24/rspb.2012.0515.full.pdf+html.

Bänziger, M., and J. Araus. 2007. "Recent Advances in Breeding Maize for Drought and Salinity Stress Tolerance." In *Advances in Molecular Breeding toward Drought and Salt Tolerant Crops*, edited by M. A. Jenks, P. M. Hasegawa, and S. M. Jain, 587–601. Netherlands: Springer.

Barnabas, B., K. Jager, and A. Feher. 2008. "The Effect of Drought and Heat Stress on Reproductive Processes in Cereals." *Plant Cell and Environment* 31 (1): 11–38.

Battisti, D. S., and R. L. Naylor. 2009. "Historical Warnings of Future Food Insecurity with Unprecedented Seasonal Heat." *Science* 323 (5911): 240–244.

Batz, F. J., W. Janssen, and K. J. Peters. 2003. "Predicting Technology Adoption to Improve Research Priority-Setting." *Agricultural Economics* 28 (2): 151–164.

Beintema, N., G. J. Stads, K. Fuglie, and P. Heisey. 2012. *ASTI Global Assessment of Agricultural R&D Spending: Developing Countries Accelerate Investment*. Washington, DC: International Food Policy Research Institute.

Belder, P., D. Rohrbach, S. Twomlow, and A. Senzanje. 2007. *Can Drip Irrigation Improve Food Security for Vulnerable Households in Zimbabwe?* Briefing Note 7. Bulawayo, Zimbabwe: International Crops Research Institute for the Semi-Arid Tropics.

Bergez, J. E., J. M. Deumier, B. Lacroix, P. Leroy, and D. Wallach. 2002. "Improving Irrigation Schedules by Using a Biophysical and a Decisional Model." *European Journal of Agronomy* 16: 123–135.

Bockus, W. W., J. A. Appel, R. L. Bowden, A. K. Fritz, B. S. Gill, T. J. Martin, R. G. Sears, D. L. Seifers, G. L. Brown-Guedira, and M. G. Eversmeyer. 2001. "Success Stories: Breeding for Wheat Disease Resistance in Kansas." *Plant Disease* 85: 453–461.

Bollinger, A., J. Magid, T. J. C. Amado, F. Skora Neto, M. de Fatima dos Santos Ribeiro, A. Calegari, R. Ralisch, and A. de Neergaard. 2006. "Taking Stock of the Brazilian 'Zero-Till Revolution': A Review of Landmark Research and Farmers' Practice." *Advances in Agronomy* 91: 1–64.

Bongiovanni, R., and J. Lowenberg-DeBoer. 2004. "Precision Agriculture and Sustainability." *Precision Agriculture* 5: 359–387.

Bottrell, D. G., and K. G. Schoenly. 2012. "Resurrecting the Ghost of Green Revolutions Past: The Brown Planthopper as a Recurring Threat to High-Yielding Rice Production in Tropical Asia." *Journal of Asia-Pacific Entomology* 15: 122–140.

Bouman, B., R. Barker, E. Humphreys, T. Phuc Tuong, G. Atlin, J. Bennett, D. Dawe, K. Dittert, A. Dobermann, T. Facon, N. Fujimoto, R. Gupta, S. Haefele, Y. Hosen, A. Ismail, D. Johnson, S. Johnson, S. Khan, L. Shan, I. Masih, Y. Matsuno, S. Pandey, S. Peng, T. M. Thiyagarajan, and R. Wassmann. 2007. "Rice: Feeding the Billions." In *Water for Food, Water for Life*, edited by D. Molden. London: Earthscan; Colombo, Sri Lanka: International Water Management Institute.

Bramley, R. G. V. 2009. "Lessons from Nearly 20 Years of Precision Agriculture Research, Development, and Adoption as a Guide to Its Appropriate Application." *Crop and Pasture Science* 60: 197–217.

Bramley, R. G. V., P. A. Hill, P. J. Thorburn, F. J. Kroon, and K. Panten. 2008. "Precision Agriculture for Improved Environmental Outcomes: Some Australian Perspectives." *Agriculture and Forestry Research* 3 (58): 161–178.

Brouwer, C., K. Prins, M. Kay, and M. Heibloem. 1988. *Irrigation Water Management: Irrigation Methods.* Training Manual 9. Rome: Food and Agriculture Organization of the United Nations.

Bruce, W. B., G. O. Edmeades, and T. C. Barker. 2002. "Molecular and Physiological Approaches to Maize Improvement for Drought Tolerance." *Journal of Experimental Botany* 53 (366): 13–25.

Burney, J., and R. Naylor. 2012. "Smallholder Irrigation as a Poverty Alleviation Tool in Africa South of the Sahara." *World Development* 40 (1): 110–123.

CAB International. n.d. Crop Protection Compendium. Oxfordshire, UK. Accessed 2012. http://www.cabi.org/CPC/.

Cairns, J. E., K. Sonder, P. H. Zaidi, N. Verhulst, G. Mahuku, R. Babu, S. K. Nair, B. Das, B. Govaerts, M. T. Vinayan, Z. Rashid, J. J. Noor, P. Devi, F. S. Vicente, and B. M. Prasanna. 2012. "Maize Production in a Changing Climate: Impacts, Adaptation, and Mitigation Strategies." *Advances in Agronomy*, Vol. 114: 1–58.

Cantero-Martinez, C., P. Angas, and J. Lampurlanes. 2003. "Growth, Yield and Water Productivity of Barley (Hordeumvulgare L.) Affected by Tillage and N Fertilization in Mediterranean Semiarid, Rainfed Conditions of Spain." *Field Crops Research* 84 (3): 341–357.

Cassman, K. G. 1999. "Ecological Intensification of Cereal Production Systems: Yield Potential, Soil Quality, and Precision Agriculture." *Proceedings of the National Academy of Sciences of the United States of America* 96 (11): 5952–5959.

———. 2007. "Editorial Response by Kenneth Cassman: Can Organic Agriculture Feed the World—Science to the Rescue?" *Renewable Agriculture and Food Systems* 22 (2): 83–84.

Chadha, R., and S. Davenport. 2011. *Agricultural Policy Reform in the BRIC Countries.* Paper for Discussion. New Delhi, India: Australian Centre for International Agricultural Research and National Council of Applied Economic Research. Available at http://www.ncaer.org/popuppages/eventdetails/e16feb2011/brics_discu_paper.pdf.

Chakraborty, S., A. V. Tiedemann, and P. S. Teng. 2000. "Climate Change: Potential Impact on Plant Diseases." *Environmental Pollution* 108 (3): 317–326.

Chivenge, P., B. Vanlauwe, and J. Six. 2011. "Does the Combined Application of Organic and Mineral Nutrient Sources Influence Maize Productivity? A Meta-Analysis." *Plant and Soil* 342 (1–2): 1–30.

Ciais, P., M. Reichstein, N. Viovy, A. Granier, J. Ogee, V. Allard, M. Aubinet, N. Buchmann, C. Bernhofer, A. Carrara, F. Chevallier, N. De Noblet, A. D. Friend, P. Friedlingstein, T. Grunwald, B. Heinesch, P. Keronen, A. Knohl, G. Krinner, D. Loustau, G. Manca, G. Matteucci, F. Miglietta, J. M. Ourcival, D. Papale, K. Pilegaard, S. Rambal, G. Seufert, J. F. Soussana, M. J. Sanz, E. D. Schulze, T. Vesala, and R. Valentini. 2005. "Europe-Wide Reduction in Primary Productivity Caused by the Heat and Drought in 2003." *Nature* 437: 529–533.

Clay, J. 2011. "Freeze the Footprint of Food." *Nature* 475: 287–289.

Connor, D. J. 2008. "Organic Agriculture Cannot Feed the World." *Field Crops Research* 106 (2): 187–190.

Cornish, G. A. 1998. "Pressurized Irrigation Technologies for Smallholders in Developing Countries: A Review." *Irrigation and Drainage Systems* 12 (3): 185–201.

Cossani, C. M., and M. P. Reynolds. 2012. "Physiological Traits for Improving Heat Tolerance in Wheat." *Plant Physiology* 160 (4): 1710–1718.

Critchley, W., and K. Siegert. 1991. *Water Harvesting: A Manual for the Design and Construction of Water Harvesting Schemes for Plant Production.* Rome: Food and Agricultural Organization of the United Nations. Available at http://www.fao.org/docrep/U3160E00.htm.

Daberkow, S. G., and W. D. McBride. 2003. "Farm and Operator Characteristics Affecting the Awareness and Adoption of Precision Agriculture Technologies in the US." *Precision Agriculture* 4: 163–177.

dePonti, T., B. Rijk, and M. K. van Ittersum. 2012. "The Crop Yield Gap between Organic and Conventional Agriculture." *Agricultural Systems* 108: 1–9.

de Rouw, A., S. Huon, B. Soulileuth, P. Jouquet, A. Pierret, O. Ribolzi, C. Valentin, E. Bourdon, and B. Chantharath. 2010. "Possibilities of Carbon and Nitrogen Sequestration under Conventional Tillage and No-Till Cover Crop Farming (Mekong Valley, Laos)." *Agriculture Ecosystems and Environment* 136 (1–2): 148–161.

Derpsch, R., and T. Friedrich. 2009. "Global Overview of Conservation Agriculture Adoption." Presented at the 4th World Congress on Conservation Agriculture, New Delhi, India, February 4–7. Available at http://www.fao.org/ag/ca/doc/global-overview-ca-adoption-derpschcomp2 .pdf.

Dhawan, B. D. 2000. "Drip Irrigation: Evaluating Returns." *Economic and Political Weekly (India),* October 14, 3775–3780.

Diagne, A. 2006. "Diffusion and Adoption of Nerica Rice Varieties in Côte D'Ivoire." *The Developing Economies* 44 (2): 208–231.

Dibb, D. W. 2000. "The Mysteries (Myths) of Nutrient Use Efficiency." *Better Crops* 84: 3–5.

Dobermann, A., and K. G. Cassman. 2005. "Cereal Area and Nitrogen Use Efficiency Are Drivers of Future Nitrogen Fertilizer Consumption." *Science in China* C 48: 745–758.

Du, B., J. Deng, W. Y. Li, and Z. X. Liao. 2000. "Comparison of Effects of Conservation and Conventional Tillage Systems on Winter Wheat Growth, Grain Yield and Soil Properties." *Journal of China Agricultural University* 5: 55–58 (in Chinese, with English abstract).

Dumanski, J., R. Peiretti, J. R. Benites, D. McGarry, and C. Pieri. 2006. "The Paradigm of Conservation Agriculture." *Proceedings of World Association of Soil and Water Tillage,* Paper P1-7, 58–64.

Ecocrop FAO (Food and Agriculture Organization of the United Nations). n.d. The Crop Environment Response database. Accessed December 6, 2012. http://Ecocrop.fao.org/ Ecocrop/srv/en/dataSheet?id=1983.

Edmeades, G. O. 2008. "Drought Tolerance in Maize: An Emerging Reality." In *Global Status of Commercialized Biotech/GM Crops: 2008,* edited by C. James, 195–217. ISAAA Briefs 39, Ithaca, NY: International Service for the Acquisition of Agri-Biotech Applications.

El-Hendawy, S. E., E. M. Hokam, and U. Schmidhalter. 2008. "Drip Irrigation Frequency: The Effects and Their Interaction with Nitrogen Fertilization on Sandy Soil Water Distribution, Maize Yield and Water Use Efficiency under Egyptian Conditions." *Journal of Agronomy and Crop Science* 194: 180–192.

Erenstein, O. 2009. *Zero Tillage in the Rice-Wheat Systems of the Indo-Gangetic Plains: A Review of Impacts and Sustainability Implications.* IFPRI Discussion Paper 00916. Washington, DC: International Food Policy Research Institute.

Fan, S., M. Torero, and D. Headey. 2011. *Urgent Actions Needed to Prevent Recurring Food Crises.* IFPRI Policy Brief 16. Washington, DC: International Food Policy Research Institute.

FAO (Food and Agriculture Organization of the United Nations). 2001. *The Economics of Conservation Agriculture.* Rome.

Ferguson, R. B., G. W. Hergert, J. S. Schepers, and C. A. Crawford. 1999. "Site-Specific Nitrogen Management of Irrigated Corn." In *Proceedings of the Fourth International Conference on Precision Agriculture,* St. Paul, MN, USA, 19–22 July 1998, part A and part B, 733–743.

Fischer, G., M. Shah, F. N. Tubiello, and H. van Velhuizen. 2005. "Socio-Economic and Climate Change Impacts on Agriculture: An Integrated Assessment." *Philosophical Transactions of the Royal Society* B 360: 2067–2083. Available at http://rstb.royalsocietypublishing.org/ content/360/1463/2067.full.

Foley, J. A., N. Ramankutty, K. A. Brauman, E. S. Cassidy, J. S. Gerber, M. Johnston, N. D. Mueller, C. O'Connell, D. K. Ray, P. C. West, C. Balzer, E. M. Bennett, S. R. Carpenter, J. Hill, C. Monfreda, S. Polasky, J. Rockström, J. Sheehan, S. Siebert, D. Tilman, and D. P. M. Zaks. 2011. "Solutions for a Cultivated Planet." *Nature* 478: 337–342.

Fountas, S., S. M. Pedersen, and S. Blackmore. 2005. "ICT in Precision Agriculture—Diffusion of Technology." In *ICT in Agriculture: Perspectives of Technological Innovation,* by E. Gelb and A.

Offer. Athens: European Federation for Information Technologies in Agriculture, Food and the Environment (EFITA); Samuel Neaman Institute for Advanced Studies in Science and Technology. http://departments.agri.huji.ac.il/economics/gelb-pedersen-5.pdf.

Fox, P., J. Rockström, and J. Barron. 2005. "Risk Analysis and Economic Viability of Water Harvesting for Supplemental Irrigation in Semi-Arid Burkina Faso and Kenya." *Agricultural Systems* 83 (3): 231–250.

Friedlander, L., A. Tal, and N. Lazarovitch. 2013. "Technical Considerations Affecting Adoption of Drip Irrigation in Africa South of the Sahara." *Agricultural Water Management* 126: 125–132.

Fuglie, K. O., P. W. Heisey, J. L. King, C. E. Pray, K. Day-Rubenstein, D. Schimmelpfennig, S. L. Wang, and R. Karmarkar-Deshmukh. 2011. *Research Investments and Market Structure in the Food Processing, Agricultural Input, and Biofuel Industries Worldwide.* ERR-130. Washington, DC: U.S. Department of Agriculture, Economic Research Service.

Garnett, T., M. C. Appleby, A. Balmford, I. J. Bateman, T. G. Benton, P. Bloomer, B. Burlingame, M. Dawkins, L. Dolan, D. Fraser, M. Herrero, I. Hoffmann, P. Smith, P. K. Thornton, C. Toulmin, S. J. Vermeulen, and H. C. J. Godfray. 2013. "Sustainable Intensification in Agriculture: Premises and Policies." *Science* 341: 33–34.

Garrett, K. A., G. A. Forbes, S. Savary, P. Skelsey, A. H. Sparks, C. Valdivia, A. H. C. van Bruggen, L. Willocquet, A. Djurle, E. Duveiller, H. Eckersten, S. Pande, C. Vera Cruz, and J. Yuen. 2011. "Complexity in Climate-Change Impacts: An Analytical Framework for Effects Mediated by Plant Disease." *Plant Pathology* 60 (1): 15–30.

Gbegbelegbe, S., D. Cammarano, S. Asseng, M. Adam, R. Robertson, J. W. Jones, K. J. Boote, O. Abdalla, T. Payne, M. Reynolds, B. Shiferaw, A. Palazzo, S. Robinson, and G. Nelson. 2012. *Promising Wheat Technologies: A Bio-Economic Modeling Approach.* Report for Global Futures for Agriculture. Nairobi: International Maize and Wheat Improvement Center.

Gebbers, R., and V. I. Adamchuck. 2010. "Precision Agriculture and Food Security." *Science* 327: 828–831.

Geng, S., J. Auburn, E. Brandstetter, and B. Li. 1988. *A Program to Simulate Meteorological Variables: Documentation for SIMMETEO.* Agronomy progress report. Davis: Univerity of California, Davis, Agricultural Experiment Station.

Gianessi, L. P. 2013. *The Potential for Worldwide Crop Production Increase Due to Adoption of Pesticides—Rice, Wheat & Maize.* Washington, DC: CropLife Foundation.

Giller, K. E., E. Witter, M. Corbeels, and P. Tittonell. 2009. "Conservation Agriculture and Smallholder Farming in Africa: The Heretics' View." *Field Crops Research* 114 (1): 23–34.

Givens, W. A., D. R. Shaw, G. R. Kruger, W. G. Johnson, S. C. Weller, B. G. Young, R. G. Wilson, M. D. K. Owen, and D. Jordan. 2009. "Survey of Tillage Trends Following the Adoption of Glyphosate-Resistant Crops." *Weed Technology* 23 (1): 150–155.

Godfray, H. C. J., J. R. Beddington, I. R. Crute, L. Haddad, D. Lawrence, J. F. Muir, J. Pretty, S. Robinson, S. M. Thomas, and C. Toulmin. 2010. "Food Security: The Challenge of Feeding 9 Billion People." *Science* 327: 812–818.

Godoy, A. C., G. A. Perez, E.C.A. Torres, L. J. Hermosillo, and J. E. L. Reyes. 2003. "Water Use, Forage Production and Water Relations in Alfalfa with Subsurface Drip Irrigation." *Agrociencia* 37: 107–115.

Goldberg, D., B. Gornat, and D. Rimon. 1976. *Drip Irrigation: Principles, Design, and Agricultural Practices.* Kfar Shmarhahu, Israel: Drip Irrigation Scientific Publications.

Gomiero, T., D. Pimentel, and M. G. Paoletti. 2011. "Environmental Impact of Different Agricultural Management Practices: Conventional vs. Organic Agriculture." *Critical Reviews in Plant Sciences* 30: 95–124.

Govaerts, B., K. D. Sayre, and J. Deckers. 2005. "Stable High Yields with Zero Tillage and Permanent Bed Planting?" *Field Crops Research* 94 (1): 33–42.

Grant, R. F., B. S. Jackson, J. R. Kiniry, and G. F. Arkin. 1989. "Water Deficit Timing Effects on Yield Components in Maize." *Agronomy Journal* 81 (1): 61–65.

Griffin, T. W., and J. Lowenberg-DeBoer. 2005. "Worldwide Adoption and Profitability of Precision Agriculture: Implications for Brazil." *Revista de Politica Agricola* 14 (4): 20–36.

Griliches, Z. 1957. "Hybrid Corn: An Exploration in the Economics of Technological Change." *Econometrica* 25 (4): 501 522.

Gruhn, P., F. Goletti, and M. Yudelman. 2000. *Integrated Nutrient Management, Soil Fertility, and Sustainable Agriculture: Current Issues and Future Challenges.* 2020 Brief 67. Washington, DC: International Food Policy Research Institute. Available at www.ifpri.org/sites/default/files/publications/brief67.pdf.

Harmel, R. D., A. L. Kenimer, S. W. Searcy, and H. A. Torbert. 2004. "Runoff Water Quality Impact of Variable Rate Sidedress Nitrogen Application." *Precision Agriculture* 5: 247–261.

Hatibu, N., K. Mutabazi, E. M. Senkondo, and A. S. K. Msangi. 2006. "Economics of Rainwater Harvesting for Crop Enterprises in Semi-Arid Areas of East Africa." *Agricultural Water Management* 80 (1–3): 74–86.

Headey, D., and S. Fan. 2010. *Reflections on the Global Food Crisis: How Did It Happen? How Has It Hurt? And How Can We Prevent the Next One?* Research Monograph 165. Washington, DC: International Food Policy Research Institute.

Heffner, L., M. E. Sorrellsa, and J.-L. Jannink. 2009. "Genomic Selection for Crop Improvement." *Crop Science* 49 (1): 1–12.

Hendrix, J. 2007. "Editorial Response by Jim Hendrix." *Renewable Agriculture and Food Systems* 22 (2): 84–85.

Hijmans, R. J. 2013. *Raster: Geographic Data Analysis and Modeling. R Package Version 2.1-66.* http://CRAN.R-project.org/package=raster.

Hijmans, R. J., S. Phillips, J. Leathwick, and J. Elith. 2013. *Dismo: Species Distribution Modeling. R Package Version 0.9-1.* http://CRAN.R-project.org/package=dism.

Hirel, B., J. Le Gouis, B. Ney, and A. Gallais. 2007. "The Challenge of Improving Nitrogen Use Efficiency in Crop Plants: Towards a More Central Role for Genetic Variability and Quantitative Genetics within Integrated Approaches." *Journal of Experimental Botany* 58 (9): 2369–2387.

Hobbs, P. R., K. Sayre, and R. Gupta. 2008. "The Role of Conservation Agriculture in Sustainable Agriculture." *Philosophical Transactions of the Royal Society* B 363 (1491): 543–555.

Hodson, D. P., and J. W. White. 2007. "Use of Spatial Analyses for Global Characterization of Wheat-Based Production Systems." *Journal of Agricultural Science* 145: 115–125.

Hoogenboom, G., J. W. Jones, C. H. Porter, P. W. Wilkens, K. J. Boote, L. A. Hunt, and G. Y. Tsuji (editors). 2010. *Decision Support System for Agrotechnology Transfer Version 4.5.* Volume 1: Overview. Honolulu: University of Hawaii.

Hoogenboom, G., J. W. Jones, P. W. Wilkens, C. H. Porter, K. J. Boote, L. A. Hunt, U. Singh, J. L. Lizaso, J. W. White, O. Uryasev, F. S. Royce, R. Ogoshi, A. J. Gijsman, G. Y. Tsuji, and J. Koo. 2012. Decision Support System for Agrotechnology Transfer (DSSAT) Version 4.5. Honolulu: University of Hawaii. CD-ROM.

Horowitz, J., R. Ebel, and K. Ueda. 2010. "'No-Till' Farming Is a Growing Practice." USDA-ERS Economic Information Bulletin EIB-70. www.ers.usda.gov/publications/eib-economic -information-bulletin/eib70.aspx#.UpN6AMRDt8F.

Hovmøller, M. S., C. K. Sorensen, S. Walter, and A. F. Justesen. 2011. "Diversity of *Puccinia striiformis* on Cereals and Grasses." *Annual Review of Phytopathology* 49: 197–217.

Hovmøller, M. S., A. H. Yahyaoui, E. A. Milus, and A. F. Justesen. 2008. "Rapid Global Spread of Two Aggressive Strains of a Wheat Rust Fungus." *Molecular Ecology* 17: 3818–3826.

Howell, T. A. 2003. "Irrigation Efficiency." In *Encyclopedia of Water Science,* edited by B. A. Stewart and T. A. Howell, 467–472. New York: Marcel Dekker.

Hughes, R. D., and G. F. Maywald. 1990. "Forecasting the Favorableness of the Australian Environment for the Russian Wheat Aphid, *Diuraphisnoxia* (Homoptera, Aphididae), and Its Potential Impact on Australian Wheat Yields." *Bulletin of Entomological Research* 80 (2): 165–175.

Hutmacher, R. B., C. J. Phene, R. M. Mead, P. Shouse, D. Clark, S. S. Vail, R. Swain, M. S. Peters, C. A. Hawk, D. Kershaw, T. Donovan, J. Jobes, and J. Fargerlund. 2001. "Subsurface Drip and Furrow Irrigation Comparison with Alfalfa in the Imperial Valley." Presented at the 31st California Alfalfa & Forage Symposium, Modesto, CA, December 11–13.

ICID (International Commission on Irrigation and Drainage). 2012. *Annual Report: 2011–2012.* New Delhi.

IDE (International Development Enterprises). n.d. *Technical Manual for IDEal Micro Irrigation Systems.* CGIAR Challenge Program on Food and Water. Lakewood, CO, USA: International Development Enterprises.

INCID (Indian National Committee on Irrigation and Drainage). 1994. *Drip Irrigation in India.* New Delhi.

IPCC (Intergovernmental Panel on Climate Change). Various years and various reports. Website. Accessed December 2013. http://ipcc.ch/.

Isika, M., G. C. M. Mutiso, and M. Muyanga. 2002. *Kitui Sand Dams and Food Security.* MVUA GHARP Newsletter 4. Nairobi: Kenya Rainwater Association.

ITC (Intermediate Technology Consultants). 2003. "Low-Cost Micro Irrigation Technologies for the Poor." Final report to the UK Department for International Department, Knowledge and Research Programme Project R7392. Rugby, UK.

Ito, M., T. Matsumoto, and M. A. Quinones. 2007. "Conservation Tillage Practice in Africa South of the Sahara: The Experience of Sasakawa Global 2000." *Crop Protection* 26: 417–423.

James, C. 2009. *Global Status of Commercialized Biotech/GM Crops: 2009.* Brief 41. Ithaca, NY: International Service for the Acquisition of Agri-Biotech Applications.

———. 2010. *Global Status of Commercialized Biotech/GM Crops: 2010.* Brief 42. Ithaca, NY: International Service for the Acquisition of Agri-Biotech Applications.

Jones, C., and P. T. Dyke. 1986. *CERES-Maize: A Simulation Model of Maize Growth and Development.* College Station: Texas A&M University Press.

Jones, J. W., G. Hoogenboom, C. H. Porter, K. J. Boote, W. D. Batchelor, L. A. Hunt, P. W. Wilkens, U. Singh, A. J. Gijsman, and J. T. Ritchie. 2003. "The DSSAT Cropping System Model." *European Journal of Agronomy* 18 (3–4): 235–265.

Kabutha, C., H. Blank, and B. Van Koppen. 2000. "Drip Kits for Smallholders in Kenya: Experience and a Way Forward." Paper presented at the 6th International Micro-Irrigation Congress on Micro-Irrigation Technology for Developing Agriculture, Cape Town, October 22–27.

Kahlown, M. A., A. Raoof, M. Zubair, and W. D. Kemper. 2007. "Water Use Efficiency and Economic Feasibility of Growing Rice and Wheat with Sprinkler Irrigation in the Indus Basin of Pakistan." *Agricultural Water Management* 87 (3): 292–298.

Kassam, A., T. Friedrich, F. Shaxson, and J. Pretty. 2009. "The Spread of Conservation Agriculture: Justification, Sustainability and Uptake." *International Journal of Agricultural Sustainability* 7 (4): 292–320.

Kayombo, B., N. Hatibu, and H. F. Mahoo. 2004. "Effect of Micro-Catchment Rainwater Harvesting on Yield of Maize in a Semi-Arid Area." Paper presented at ISCO 2004: 13th International Soil Conservation Organization Conference, Brisbane, Australia, July 1–4.

Khan, M. A., M. Iqbal, M. Jameel, W. Nazeer, S. Shakir, M. T. Aslam, and B. Iqbal. 2011. "Potentials of Molecular Based Breeding to Enhance Drought Tolerance in Wheat (*Triticum aestivum* L.)." *African Journal of Biotechnology* 10 (55): 11340–11344.

Kirchmann, H., T. Kaetterer, and L. Bergstroem. 2008. "Nutrient Supply in Organic Agriculture—Plant Availability, Sources and Recycling." In *Organic Crop Production—Ambitions and Limitations,* edited by H. Kirchmann and L. Bergstroem, 89–116. Dordrecht: Springer.

Kocmankova, E., M. Trnka, J. Eitzinger, H. Formayer, M. Dubrovsky, D. Semeradova, Z. Zalud, J. Juroch, and M. Mozny. 2010. "Estimating the Impact of Climate Change on the Occurrence of Selected Pests in the Central European Region." *Climate Research* 44 (1): 95–105.

Koo, J., and J. Dimes. 2010, HC27 Generic Soil Profile database, Version V1. Washington, DC: International Food Policy Research Institute. http://hdl.handle.net/1902.1/20299.

Krupinsky, J. M., K. L. Bailey, M. P. McMullen, B. D. Gossen, and T. K. Turkington. 2002. "Managing Plant Disease Risk in Diversified Cropping Systems." *Agronomy Journal* 94 (2): 198–209.

Kulkarni, S. A., F. B. Reinders, and F. Ligetvari. 2006. "Global Scenario of Sprinklers in Micro-Irrigated Areas." Paper presented at the 7th International Micro Irrigation Congress, Kuala Lumpur, September 13–15.

Kumar, A. 2011. "Bangladesh Releases Two Drought Tolerant Varieties." *STRASA News* 4 (3): 2.

Kumar, A., and M. Frio. 2011. "Drought Problems No More." *STRASA News* 4 (3): 2–5.

Lecina, S., D. Isidoro, E. Playan, and R. Aragues. 2010. "Irrigation Modernization and Water Conservation in Spain: The Case of Riegos del Alto Aragon." *Agricultural Water Management* 97 (10): 1663–1675.

Legg, D., and M. Cohen. 2003. "Computer-Simulated Evaluation of Insect Phenology Models Developed in the Tropics: *Chilo suppressalis* (Walker) as an Example." *International Journal of Pest Management* 49 (3): 215–224.

Lemoalle, J., and D. de Condappa. 2012. "Farming Systems and Food Production in the Volta Basin." In *Water, Food and Poverty in River Basins: Defining the Limits,* edited by M. Fisher and S. Cook, 192–217. London: Routledge.

Lerch, R. N., N. R. Kitchen, R. J. Kremer, W. W. Donald, E. E. Alberts, E. J. Sadler, K. A. Sudduth, D. B. Myers, and F. Ghidey. 2005. "Development of a Conservation-Oriented Precision Agriculture System: Water and Soil Quality Assessment." *Journal of Soil and Water Conservation* 60 (6): 411–421.

Liniger, H. P., R. Mekdaschi Studer, C. Hauert, and M. Gurtner. 2011. *Sustainable Land Management in Practice—Guidelines and Best Practices for Africa South of the Sahara*. Rome: TerrAfrica, World Overview of Conservation Approaches and Technologies, and Food and Agriculture Organization of the United Nations.

Lobell, D. B., M. Banziger, C. Magorokosho, and B. Vivek. 2011. "Nonlinear Heat Effects on African Maize as Evidenced by Historical Yield Trials." *Nature Climate Change* 1 (1): 42–45.

Lobell, D. B., and M. B. Burke. 2009. *Climate Change and Food Security: Adapting Agriculture to a Warmer World*. New York: Springer.

———. 2010. "The Use of Statistical Models to Predict Crop Yield Responses to Climate Change." *Agricultural and Forest Meteorology* 150 (11): 1443–1452.

Lobell, D. B., A. Sibley, and J. I. Ortiz-Monasterio. 2012. "Extreme Heat Effects on Wheat Senescence in India." *Nature Climate Change* 2 (3): 186–189.

Lopes, M. S., J. L. Araus, P. D. R. van Heerden, and C. H. Foyer. 2011. "Enhancing Drought Tolerance in C (4) Crops." *Journal of Experimental Botany* 62 (9): 3135–3153.

Lowenberg-DeBoer, J., and T. W. Griffin. 2006. *Potential for Precision Agriculture Adoption in Brazil*. Site-Specific Management Center Newsletter, Purdue University.

MA and WRI (Millennium Ecosystem Assessment and World Resources Institute). 2005. *Ecosystems and Human Well-Being: General Synthesis*. Washington, DC: Island Press.

Mackill, D. J., U. S. Singh, M. J. Thomson, E. Septiningsih, and A. Kumar. 2010. "Technological Opportunities for Developing and Deploying Improved Germplasm for Key Target Traits." In *Rice in the Global Economy: Strategic Research and Policy Issues for Food Security*, edited by S. Pandey, D. Byerlee, D. Dawe, A. Dobermann, S. Mohanty, S. Rozelle, and B. Hardy, 433–448. Los Baños, the Philippines: International Rice Research Institute.

Maisiri, N., A. Senzanje, J. Rockström, and S. J. Twomlow. 2005. "On Farm Evaluation of the Effect of Low Cost Drip Irrigation on Water and Crop Productivity Compared to Conventional Surface Irrigation Systems." *Physics and Chemistry of the Earth* 30: 783–791.

Margosian, M. L., K. A. Garrett, J. M. S. Hutchinson, and K. A. With. 2009. "Connectivity of the American Agricultural Landscape: Assessing the National Risk of Crop Pest and Disease Spread." *BioScience* 59 (2): 141–151.

Margulis, S. 2010. *Economics of Adaptation to Climate Change: Synthesis Report*. Washington, DC: World Bank.

Mateete, B., S. Nteranya, and P. L. Woomer. 2010. "Restoring Soil Fertility in Africa South of the Sahara." *Advances in Agronomy* 108: 183–236.

McBratney, A., B. Whelan, and T. Ancev. 2005. "Future Directions of Precision Agriculture." *Precision Agriculture* 6: 7–23.

Messmer, R., Y. Fracheboud, M. Banziger, P. Stamp, and J. M. Ribaut. 2011. "Drought Stress and Tropical Maize: QTLs for Leaf Greenness, Plant Senescence, and Root Capacitance." *Field Crops Research* 124 (1): 93–103.

Milus, E. A., E. Seyran, and R. McNew. 2006. "Aggressiveness of *Puccinia striiformis* f. sptritici Isolates in the South-Central United States." *Plant Disease* 90 (7): 847–852.

Molden, D. 2007. *Water for Food, Water for Life: A Comprehensive Assessment of Water Management in Agriculture.* London: Earthscan; Colombo, Sri Lanka: International Water Management Institute.

Möller, M., and E. K. Weatherhead. 2007. "Evaluating Drip Irrigation in Commercial Tea Production in Tanzania." *Irrigation Drainage Systems* 21: 17–34.

Moyo, R., D. Love, M. Mul, W. Mupangwa, and S. Twomlow. 2006. "Impact and Sustainability of Low-Head Drip Irrigation Kits in the Semi-Arid Gwanda and Beitbridge Districts, Mzingwane Catchment, Limpopo Basin, Zimbabwe." *Physics and Chemistry of the Earth* 31: 885–892. doi:10.1016/j.pce.2006.08.020.

Mrabet, R. 2008. *No-Tillage Systems for Sustainable Dryland Agriculture in Morocco.* Fanigraph Edition. Paris: National Institute for Agricultural Research (INRA).

Mugwe, J., D. Mugendi, M. Mucheru-Muna, R. Merckx, J. Chianu, and B. Vanlauwe. 2009. "Determinants of the Decision to Adopt Integrated Soil Fertility Management Practices by Smallholder Farmers in the Central Highlands of Kenya." *Experimental Agriculture* 45 (1): 61–75.

Nachtergaele, F., H. van Velthuizen, L. Verelst, and D. Wiberg. 2012. Harmonized World Soil Database, version 1.2. Rome: Food and Agriculture Organization of the United Nations; Laxenburg, Austria: International Institute for Applied Systems Analysis.

Narayanamoorthy, A. 1996. *Evaluation of Drip Irrigation System in Maharashtra.* Mimeograph Series 42, Agro Economic Research Centre. Pune, India: Gokhale Institute of Politics and Economics.

———. 1997. "Economic Viability of Drip Irrigation: An Empirical Analysis from Maharashtra." *Indian Journal of Agricultural Economics* 52 (4): 728–739.

———. 2008. *Economics of Drip Irrigated Cotton: A Synthesis of Four Case Studies.* Conference Paper H042297. Colombo, Sri Lanka: International Water Management Institute.

Naseem, A., D. J. Spielman, and S. W. Omamo. 2010. "Private-Sector Investment in R&D: A Review of Policy Options to Promote Its Growth in Developing-Country Agriculture." *Agribusiness* 26 (1): 143–173.

Nelson, G. C., M. W. Rosegrant, J. Koo, R. Robertson, T. Sulser, T. Zhu, C. Ringler, S. Msangi, A. Palazzo, M. Batka, M. Magalhaes, R. Valmonte-Santos, M. Ewing, and D. Lee. 2009. *Climate*

Change: Impact on Agriculture and Costs of Adaptation. IFPRI Food Policy Report. Washington, DC: International Food Policy Research Institute.

Nelson, G. C., M. W. Rosegrant, A. Palazzo, I. Gray, C. Ingersoll, R. Robertson, S. Tokgoz, T. Zhu, T. B. Sulser, C. Ringler, S. Msangi, and L. You. 2010. *Food Security, Farming, and Climate Change to 2050: Scenarios, Results, and Policy Options.* Washington, DC: International Food Policy Research Institute.

Nelson, G. C., D. van der Mensbrugghe, H. Ahammad, E. Blanc, K. Calvin, T. Hasegawa, P. Havlik, E. Heyhoe, P. Kyle, H. Lotze-Campen, M. von Lampe, D. Mason d'Croz, H. van Meijl, C. Müller, J. Reilly, R. Robertson, R. D. Sands, C. Schmitz, A. Tabeau, K. Takahashi, H. Valin, and D. Willenbockel. 2013. "Agriculture and Climate Change in Global Scenarios: Why Don't the Models Agree." *Agricultural Economics.* doi: 10.1111/agec.12091.

Neogi, M. G., and R. M. Baltazar. 2011. "Drought Tolerant Rice Eases the Effects of Climate Change." *STRASA News* 4 (1–2): 8–9.

Ngigi, S. N. 2003. "What Is the Limit of Up-Scaling Rainwater Harvesting in a River Basin?" *Physics and Chemistry of the Earth* 28 (20–27): 943–956.

Ngigi, S. N., H. H. G. Savenije, J. Rockström, and C. K. Gachene. 2005. "Hydro-Economic Evaluation of Rainwater Harvesting and Management Technologies: Farmers' Investment Options and Risks in Semi-Arid Laikipia District of Kenya." *Physics and Chemistry of the Earth* 30 (11–16): 772–782.

Oerke, E. C. 2006. "Crop Losses to Pests." *Journal of Agricultural Science* 144: 31–43.

Oerke, E. C., and H. W. Dehne. 2004. "Safeguarding Production—Losses in Major Crops and the Role of Crop Protection." *Crop Protection* 23: 275–285.

Oerke, E. C., H. W. Dehne, F. Schönbeck, and A. Weber. 1994. *Crop Production and Crop Protection.* Amsterdam: Elsevier Science.

Ogbonnaya, F. C., G. Y. Ye, R. Trethowan, F. Dreccer, D. Lush, J. Shepperd, and M. Van Ginkel. 2007. "Yield of Synthetic Backcross-Derived Lines in Rainfed Environments of Australia." *Euphytica* 157 (3): 321–336.

Ortiz, R., H. J. Braun, J. Crossa, J. H. Crouch, G. Davenport, J. Dixon, S. Dreisigacker, E. Duveiller, Z. H. He, J. Huerta, A. K. Joshi, M. Kishii, P. Kosina, Y. Manes, M. Mezzalama, A. Morgounov, J. Murakami, J. Nicol, G. O. Ferrara, J. I. Ortiz-Monasterio, T. S. Payne, R. J. Pena, M. P. Reynolds, K. D. Sayre, R. C. Sharma, R. P. Singh, J. K. Wang, M. Warburton, H. X. Wu, and M. Iwanaga. 2008. "Wheat Genetic Resources Enhancement by the International Maize and Wheat Improvement Center (CIMMYT)." *Genetic Resources and Crop Evolution* 55 (7): 1095–1140.

Oweis, T., and A. Hachum. 2009. "Water Harvesting for Improved Rainfed Agriculture in the Dry Environments." In *Rainfed Agriculture: Unlocking the Potential,* edited by S. P.

Wani, J. Rockström, and T. Oweis, 164–181. Wallingford, UK, and Cambridge, MA: CABI International.

Pardey, P. G., and P. L. Pingali. 2010. "Reassessing International Agricultural Research." Report prepared for the Global Conference on Agricultural Research for Development, Montpellier, France, 28–31 March.

Pathak, R. R., S. Lochab, and N. Raghuram. 2011. "Improving Plant Nitrogen-Use Efficiency." *Comprehensive Biotechnology* 4: 209–218.

Perry, C., P. Steduto, R. G. Allen, and C. M. Burt. 2009. "Increasing Productivity in Irrigated Agriculture: Agronomic Constraints and Hydrological Realities." *Agricultural Water Management* 96 (11): 1517–1524.

Pfeiffer, W. H., R. M. Trethowan, M. Van Ginkel, M. I. Ortiz, and S. Rajaram. 2005. "Breeding for Abiotic Stress Tolerance in Wheat." In *Abiotic Stresses: Plant Resistance through Breeding and Molecular Approaches,* edited by M. Ashraf and P. J. C. Harris, 401–490. New York: Haworth Press.

Pieri, C., G. Evers, J. Landers, P. O'Connell, and E. Terry. 2002. *No-Till Farming for Sustainable Rural Development.* Agriculture and Rural Development Working Paper. Washington, DC: World Bank.

Pimentel, D., P. Hepperly, J. Hanson, D. Douds, and R. Seidel. 2005. "Environmental, Energetic, and Economic Comparisons of Organic and Conventional Farming Systems." *BioScience* 55 (7): 573–582.

Place, F., C. B. Barrett, H. A. Freeman, J. J. Ramisch, and B. Vanlauwe. 2003. "Prospects for Integrated Soil Fertility Management Using Organic and Inorganic Inputs: Evidence from Smallholder African Agricultural Systems." *Food Policy* 28 (4): 365–378.

Porter, C. H., J. W. Jones, S. Adiku, A. J. Gijsman, O. Gargiulo, and J. B. Naab. 2010. "Modeling Organic Carbon and Carbon-Mediated Soil Processes in DSSAT v4. 5." *Operational Research* 10 (3): 247–278.

Potter, P., N. Ramankutty, E. M. Bennett, and S. D. Donner. 2010. "Characterizing the Spatial Patterns of Global Fertilizer Application and Manure Production." *Earth Interactions* 14: 1–22.

Prasanna, B. M., J. Cairns, D. Makumbi, B. Shiferaw, and S. Gbegbelegbe. 2011. *Promising Technologies for Improving Productivity of Maize.* Mexico, D.F., Mexico: CIMMYT and Global Futures for Agriculture.

Pray, C. E., and K. O. Fuglie. 2001. *Private Investment in Agricultural Research and International Technology Transfer in Asia.* Agricultural Economics Report 805. Washington, DC: US Department of Agriculture Economic Research Service.

Pray, C. E., K. O. Fuglie, and D. K. N. Johnson. 2007. "Private Agricultural Research." In *Handbook of Agricultural Economics,* Volume 3, edited by R. Evenson and P. Pingali. Amsterdam: Elsevier.

Pretty, J., C. Toulmin, and S. Williams. 2011. "Sustainable Intensification in African Agriculture." *International Journal of Agricultural Sustainability* 9 (1): 5–24.

R Core Team. 2013. *R: A Language and Environment for Statistical Computing.* Vienna: R Foundation for Statistical Computing. Available at http://www.R-project.org/.

Reganold, J., L. Elliott, and Y. Unger. 1987. "Long-Term Effects of Organic and Conventional Farming on Soil Erosion." *Nature* 330: 370–372.

Reynolds, M., F. Dreccer, and R. Trethowan. 2007. "Drought-Adaptive Traits Derived from Wheat Wild Relatives and Landraces." *Journal of Experimental Botany* 58 (2): 177–186.

Ringler, C., A. Bhaduri, and R. Lawford. 2013. "The Nexus across Water, Energy, Land and Food (WELF): Potential for Improved Resource Use Efficiency?" *Current Opinion in Environmental Sustainability* 5 (6): 617–624.

Robert, P. C. 2002. "Precision Agriculture: A Challenge for Crop Nutrition Management." *Plant and Soil* 247 (1): 143–149.

Roberts, T. L. 2008. "Improving Nutrient Use Efficiency." *Turkish Journal of Agriculture and Forestry* 32 (3): 177–182.

Rockström, J., J. Barron, and P. Fox. 2002. "Rainwater Management for Increased Productivity among Small-Holder Farmers in Drought Prone Environments." *Physics and Chemistry of the Earth* 27 (11–22): 949–959.

Rockström J., W. Steffen, K. Noone, Å. Persson, F. S. Chapin III, E. Lambin, T. M. Lenton, M. Scheffer, C. Folke, II. Schellnhuber, B. Nykvist, C. A. De Wit, T. Hughes, S. van der Leeuw, H. Rodhe, S. Sörlin, P. K. Snyder, R. Costanza, U. Svedin, M. Falkenmark, L. Karlberg, R. W. Corell, V. J. Fabry, J. Hansen, B. Walker, D. Liverman, K. Richardson, P. Crutzen, and J. Foley. 2009. "Planetary Boundaries: Exploring the Safe Operating Space for Humanity." *Ecology and Society* 14(2): 32. www.ecologyandsociety.org/vol14/iss2/art32/.

Rosegrant, M. W., X. Cai, and S. Cline. 2002. *World Water and Food to 2025: Dealing with Scarcity.* Washington, DC: International Food Policy Research Institute.

Rosegrant, M. W., X. Cai, S. Cline, and N. Nakagawa. 2002. *The Role of Rainfed Agriculture in the Future of Global Food Production.* EPTD Discussion Paper 90. Washington, DC: International Food Policy Research Institute.

Rosegrant, M. W., M. Fernandez, and A. Sinha. 2009. "Looking into the Future for Agriculture and AKST." In *International Assessment of Agricultural Knowledge, Science and Technology for Development (IAASTD): Global Report,* edited by B. D. McIntyre, H. R. Herren, J. Wakhungu, and R. T. Watson, 307–376. Washington, DC: Island Press.

Rosegrant, M. W., and IMPACT Development Team. 2012. *International Model for Policy Analysis of Agricultural Commodities and Trade (IMPACT): Model Description.* Washington, DC: International Food Policy Research Institute.

Rosegrant, M. W., M. S. Paisner, and S. Meijer. 2003. "The Future of Cereal Yields and Prices: Implications for Research and Policy." *Journal of Crop Production* 9 (1/2): 661–690.

Rosegrant, M. W., M. Paisner, S. Meijer, and J. Witcover. 2001. *Global Food Projections to 2020: Emerging Trends and Alternative Futures*. Washington, DC: International Food Policy Research Institute.

Rosegrant, M. W., S. Tokgoz, and P. Bhandary. 2013. "The New Normal? A Tighter Global Agricultural Supply and Demand Relation and Its Implications for Food Security." *American Journal of Agricultural Economics* 95 (2): 303–309.

Royal Society. 2009. *Reaping the Benefits: Science and the Sustainable Intensification of Global Agriculture*. London.

Sanchez-Giron, V., A. Serrano, M. Suarez, J. L. Hernanz, and L. Navarrete. 2007. "Economics of Reduced Tillage for Cereal and Legume Production on Rainfed Farm Enterprises of Different Sizes in Semiarid Conditions." *Soil and Tillage Research* 95 (1–2): 149–160.

Savary, S., P. S. Teng, L. Willocquet, and F. W. Nutter Jr. 2006. "Quantification and Modeling of Crop Losses: A Review of Purposes." *Annial Review of Phytopathology* 44: 89–112.

Savary, S., L. Willocquet, F. A. Elazegui, P. S. Teng, P. V. Du, D. Zhu, Q. Tang, X. Lin, H. M. Singh, and R. K. Srivastava. 2000. "Rice Pest Constraints in Tropical Asia: Characterization of Injury Profiles in Relation to Production Situations." *Plant Diseases* 84: 341–356.

SDSN (Sustainable Development Solutions Network). 2013. *Sustainable Agriculture and Food Systems*. Technical report for the post-2015 development agenda. Available at http://unsdsn .org/files/2013/07/130724-TG-7-Report-WEB.pdf.

Serraj, R., K. L. McNally, I. Slamet-Loedin, A. Kohli, S. M. Haefele, G. Atlin, and A. Kumar. 2011. "Drought Resistance Improvement in Rice: An Integrated Genetic and Resource Management Strategy." *Plant Production Science* 14 (1): 1–14.

Seufert, V., N. Ramankutty, and J. A. Foley. 2012. "Comparing the Yields of Organic and Conventional Agriculture." *Nature* 485: 229–234.

Shah, F., J. Huang, K. Cui, L. Nie, T. Shah, C. Chen, and K. Wang. 2011. "Impact of High-Temperature Stress on Rice Plant and Its Traits Related to Tolerance." *Journal of Agricultural Science* 149: 545–556.

Sharma, S. K. 1984. *Principles and Practices of Irrigation Engineering*. New York and Oxford: IBH Publications.

Shaw, M. W., and T. M. Osborne. 2011. "Geographic Distribution of Plant Pathogens in Response to Climate Change." *Plant Pathology* 60 (1): 31–43.

Silva, C. B., S. M. LeiteRibeiro do Vale, F. A. C. Pinto, C. A. S. Müller, and A. D. Moura. 2007. "The Economic Feasibility of Precision Agriculture in MatoGrosso do Sul State, Brazil: A Case Study." *Precision Agriculture* 8: 255–265.

Singh, N., R. Agarwal, A. Awasthi, P. K. Gupta, and S. K. Mittal. 2010. "Characterization of Atmospheric Aerosols for Organic Tarry Matter and Combustible Matter during Crop Residue Burning and Non-Crop Residue Burning Months in the Northwestern Region of India." *Atmospheric Environment* 44 (10): 1292–1300.

Singh, R. P., D. P. Hodson, J. Huerta-Espino, Y. Jin, S. Bhavani, P. Njau, S. Herrera-Foessel, P. K. Singh, S. Singh, and V. Govindan. 2011. "The Emergence of Ug99 Races of the Stem Rust Fungus Is a Threat to World Wheat Production." *Annual Review of Phytopathology* 49: 465–481.

Singh, U., and P. K. Thornton. 1992. "Using Crop Models for Sustainability and Environmental-Quality Assessment." *Outlook on Agriculture* 21 (3): 209–218.

Sivanappan, R. K. 1994. "Prospects of Micro-Irrigation in India." *Irrigation and Drainage Systems* 8: 49–58.

Smith, L., and L. Haddad 2000. "Explaining Child Malnutrition in Developing Countries: A Cross-Country Analysis." Washington, DC: International Food Policy Research Institute.

Smith, P. 2013. "Delivering Food Security without Increasing Pressure on Land." *Global Food Security* 2: 18–23.

Strange, R. N., and P. R. Scott. 2005. "Plant Disease: A Threat to Global Food Security." *Annual Review of Phytopathology* 43: 83–116.

Subash, N., and H. S. R. Mohan. 2012. "Evaluation of the Impact of Climatic Trends and Variability in Rice-Wheat System Productivity Using Cropping System Model DSSAT over the Indo-Gangetic Plains of India." *Agricultural and Forest Meteorology* 164: 71–81.

Swinton, S. M., and J. Lowenberg-DeBoer. 2001. *Global Adoption of Precision Agriculture Technologies: Who, When and Why?* Ann Arbor: Department of Agricultural Economics, Michigan State University. Accessed August 2011. https://www.msu.edu/user/swintons/D7_8_SwintonECPA01.pdf.

Tollefson, J. 2011. "Drought-Tolerant Maize Gets US Debut." *Nature* 469: 144.

Tuomisto, H. L., I. D. Hodge, P. Riordan, and D. W. Macdonald. 2012. "Does Organic Farming Reduce Environmental Impacts? A Meta-Analysis of European Research." *Journal of Environmental Management* 112: 309–320.

UN (United Nations). 2011. *World Population Prospects: The 2010 Revision, Highlights and Advance Tables.* Working Paper ESA/P/WP.220. New York: United Nations, Department of Economic and Social Affairs, Population Division.

Upadhyay, B., M. Samad, and M. Giordano. 2005. *Livelihoods and Gender Roles in Drip-Irrigation Technology: A Case of Nepal.* IWMI Working Paper 87. Colombo, Sri Lanka: International Water Management Institute.

Van der Kooij, S. M. Zwarteveen, M. Boesveld, and M. Kuper. 2013. "The Efficiency of Drip Irrigation Unpacked." *Agricultural Water Management* 123: 103–110.

Vanlauwe, B., J. Kihara, P. Chivenge, P. Pypers, R. Coe, and J. Six. 2011. "Agronomic Use Efficiency of N Fertilizer in Maize-Based Systems in Africa South of the Sahara within the Context of Integrated Soil Fertility Management." *Plant and Soil* 339 (1–2): 35–50.

Verulkar, S. B., N. P. Mandal, J. L. Dwivedi, B. N. Singh, P. K. Sinha, R. N. Mahato, P. Dongre, O. N. Singh, L. K. Bose, P. Swain, S. Robin, R. Chandrababu, S. Senthil, A. Jain, H. E. Shashidhar, S. Hittalmani, C. V. Cruz, T. Paris, A. Raman, S. Haefele, R. Serraj, G. Atlin, and A. Kumar. 2010 "Breeding Resilient and Productive Genotypes Adapted to Drought-Prone Rainfed Ecosystem of India." *Field Crops Research* 117 (2–3): 197–208.

Vohland, K., and B. Barry. 2009. "A Review of in situ Rainwater Harvesting (RWH) Practices Modifying Landscape Functions in African Drylands." *Agriculture Ecosystems and Environment* 131 (3–4): 119–127.

von Grebmer, K., D. Headey, C. Béné, L. Haddad, T. Olofinbiyi, D. Wiesmann, H. Fritschel, S. Yin, Y. Yohannes, C. Foley, C. von Oppeln, and B. Iseli. 2013. *2013 Global Hunger Index: The Challenge of Hunger: Building Resilience to Achieve Food and Nutrition Security*. Bonn: Welthungerhilfe; Washington, DC: International Food Policy Research Institute; Dublin: Concern Worldwide.

Wang, F. H., Z. H. He, K. Sayre, S. D. Li, J. S. Si, B. Feng, and L. G. Kong. 2009. "Wheat Cropping Systems and Technologies in China." *Field Crops Research* 111 (3): 181–188.

Wang, J., J. Huang, L. Zhang, Q. Huang, and S. Rozelle. 2010. "Water Governance and Water Use Efficiency: The Five Principles of WUA Management and Performance in China." *Journal of the American Water Resources Association* 46 (4): 665–685.

Ward, F. A., and M. Pulido-Velazquez. 2008. "Water Conservation in Irrigation Can Increase Water Use." *Proceedings of the National Academy of Sciences, USA* 105 (47): 18215–18220.

Way, M. J., and K. L. Heong. 1994. "The Role of Biodiversity in the Dynamics and Management of Insect Pests of Tropical Irrigated Rice—A Review." *Bulletin of Entomological Research* 84: 567–587.

Willer, H., and L. Kilcher. 2011. *The World of Organic Agriculture—Statistics and Emerging Trends 2011*. Bonn: International Federation of Organic Agricultural Movements; Frick, Switzerland: Research Institute of Organic Agriculture.

Willocquet, L., S. Savary, L. Fernandez, F. A. Elazegui, N. Castilla, D. Zhu, Q. Tang, S. Huang, X. Lin, H. M. Singh, and R. K. Srivastsava. 2002. "Structure and Validation of RICEPEST, a Production Situation-Driven, Crop Growth Model Simulating Rice Yield Response to Multiple Pest Injuries for Tropical Asia." *Ecological Modeling* 153: 247–268.

Wisser, D., S. Frolking, E. M. Douglas, B. M. Fekete, A. H. Schumann, and C. J. Vorosmarty. 2010. "The Significance of Local Water Resources Captured in Small Reservoirs for Crop Production—A Global-Scale Analysis." *Journal of Hydrology* 384 (3–4): 264–275.

World Bank. n.d. *Pest Management Guidebook.* Accessed November 26, 2013. http://web.worldbank
.org/WBSITE/EXTERNAL/TOPICS/EXTARD/EXTPESTMGMT/0,,contentMDK:
20631451~menuPK:1605318~pagePK:64168445~piPK:64168309~theSitePK:584320,00
.html.

Xu, Y., and J. H. Crouch. 2008. "Marker-Assisted Selection in Plant Breeding: From Publications to
Practice." *Crop Science* 48 (2): 391–407.

Yamamura, K., M. Yokozawa, M. Nishimori, Y. Ueda, and T. Yokosuka. 2006. "How to
Analyze Long-Term Insect Population Dynamics under Climate Change: 50-Year Data on
Three Insect Pests in Paddy Fields." *Population Ecology* 48: 31–48.

You, L. Z., S. Crespo, Z. Guo, J. Koo, W. Ojo, K. Sebastian, M. T. Tenorio, S. Wood, and U. Wood-
Sichra. n.d. "Spatial Production Allocation Model (SPAM) 2000 Version 3, Release 2." Accessed
June 2012. http://MapSPAM.info.

You, L. Z., S. Wood, and U. Wood-Sichra. 2006. "Generating Global Crop Maps: From Census to
Grid." Paper selected for the International Association of Agricultural Economists Annual
Conference, Gold Coast, Australia, August.

Yu, W., and H. G. Jensen. 2010. "China's Agricultural Policy Transition: Impacts of Recent Reforms
and Future Scenarios." *Journal of Agricultural Economics* 61 (2): 343–368.

Yuan, T., F. M. Li, and P. H. Liu. 2003. "Economic Analysis of Rainwater Harvesting and Irrigation
Methods, with an Example from China." *Agricultural Water Management* 60 (3): 217–226.

Zentner, R. P., G. P. Lafond, D. A. Derksen, C. N. Nagy, D. D. Wall, and W. E. May. 2004. "Effects of
Tillage Method and Crop Rotation on Non-Renewable Energy Use Efficiency for a Thin Black
Chernozem in the Canadian Prairies." *Soil and Tillage Research* 77 (2): 125–136.

Ziska, L. H. 2002. "Influence of Rising Atmospheric CO2 since 1900 on Early Growth and
Photosynthetic Response of a Noxious Invasive Weed, Canada Thistle (Cirsiumarvense)."
Functional Plant Biology 29 (12): 1387–1392.

Zou, X., Y. Li, R. Cremades, Q. Gao, Y. Wan, and X. Qin. 2013. "Cost-Effectiveness Analysis of
Water-Saving Irrigation Technologies Based on Climate Change Response: A Case Study of
China." *Agricultural Water Management* 129: 9–20.

Authors

Mark W. Rosegrant (m.rosegrant@cgiar.org) is the director of the
Environment and Production Technology Division, International Food Policy
Research Institute (IFPRI). With a PhD in public policy from the University
of Michigan, he has extensive experience in research and policy analysis in
agriculture and economic development, with an emphasis on water resources
and other critical natural resource and agricultural policy issues as they affect
food security, rural livelihoods, and environmental sustainability. He currently
directs research on climate change, water resources, sustainable land management, genetic resources and biotechnology, and agriculture and energy. He is
the author or editor of 12 books and more than 100 refereed papers in agricultural economics, water resources, and food policy analysis. Rosegrant
has won numerous awards, and is a Fellow of the American Association for
the Advancement of Science and a Fellow of the Agricultural and Applied
Economics Association. His publications include *Climate Change: Impact
on Agriculture and Costs of Adaptation,* co-authored with G. Nelson, J. Koo,
R. Robertson, T. Sulser, T. Zhu, C. Ringler, S. Msangi, A. Palazzo, M. Batka,
M. Magalhaes, R. Valmonte-Santos, M. Ewing, and D. Lee (IFPRI); *Building
Climate Resilience in the Agriculture Sector* (Asian Development Bank and
IFPRI); and *Food Security, Farming, and Climate Change to 2050,* co-
authored with G. Nelson, A. Palazzo, I. Gray, C. Ingersoll, R. Robertson,
S. Tokgoz, T. Zhu, T. B. Sulser, C. Ringler, S. Msangi, and L. You (IFPRI).

Jawoo Koo (j.koo@cgiar.org) is a research fellow at IFPRI. He joined the
organization as a postdoctoral fellow in the Environment and Production
Technology Division. He holds a bachelor's degree in agricultural biology
from Korea University, Seoul, Korea, and a master's and PhD in agricultural
and biological engineering from the University of Florida. For his dissertation,
he worked for the Soil Management Collaborative Research Support Program,
funded by USAID, in collaboration with the University of Ghana and the

Savannah Agriculture Research Institute, to estimate soil carbon sequestration potential in smallholders' farming systems in Ghana using crop systems models and field surveys. His publications include *Climate Change: Impact on Agriculture and Costs of Adaptation,* co-authored with G. Nelson, M. W. Rosegrant, R. Robertson, T. Sulser, T. Zhu, C. Ringler, S. Msangi, A. Palazzo, M. Batka, M. Magalhaes, R. Valmonte-Santos, M. Ewing, and D. Lee (IFPRI); and *Building Climate Resilience in the Agriculture Sector* (Asian Development Bank and IFPRI).

Nicola Cenacchi (n.cenacchi@cgiar.org) is a research analyst in the Environment and Production Technology Division, IFPRI, where he works on projects related to agricultural technologies, climate change, and adaptation to extreme weather events. Previously he was part of the core team for the World Bank *2010 World Development Report on Climate Change and Development.* While working for the World Bank, Cenacchi also co-authored a chapter focusing on land and water resources, coastal areas, and biodiversity in the book *Adapting to Climate Change in Eastern Europe and Central Asia* (World Bank). He holds a Laurea in biological sciences from the University of Bologna (Italy), and a master's degree in environmental technology from Imperial College London, with a specialization in ecological management of natural resources.

Claudia Ringler (c.ringler@cgiar.org) was appointed deputy division director of the Environment and Production Technology Division, IFPRI, in 2011. From 1996 until her current appointment, she served in various other research positions in that division. She currently co-leads IFPRI's water research program and is also a basin theme leader in the CGIAR Research Program on Water, Land, and Ecosystems. Ringler received her PhD in agricultural economics from the Center for Development Research, Bonn University, Germany, and her master's degree in international and development economics from Yale University. Her research interests are water resources management (in particular, river basin modeling for policy analysis and agricultural research) and natural resource policy focused on sustainable agricultural productivity growth. Over the past several years, she has also undertaken research on the impacts of global warming for developing-country agriculture and on appropriate adaptation and mitigation options. Her publications include *Climate Change: Impact on Agriculture and Costs of Adaptation,* co-authored with G. Nelson, M. W. Rosegrant, J. Koo, R. Robertson, T. Sulser, T. Zhu, S. Msangi, A. Palazzo, M. Batka, M. Magalhaes, R. Valmonte-Santos, M. Ewing, and D. Lee (IFPRI); *Building Climate Resilience in the Agriculture Sector* (Asian Development Bank and IFPRI); and *Food Security, Farming, and Climate Change to 2050,* co-authored with G. Nelson, M. W. Rosegrant, A. Palazzo, I. Gray, C. Ingersoll, R. Robertson, S. Tokgoz, T. Zhu, T. B. Sulser, S. Msangi, and L. You (IFPRI).

Richard Robertson (r.robertson@cgiar.org) is a research fellow at IFPRI. He joined IFPRI in 2008. Prior to that he completed work related to environmental and water issues at the University of Illinois at Urbana-Champaign in the Civil and Environmental Engineering as well as the Agricultural and Consumer Economics departments. His PhD research (University of Illinois) concerned empirically modeling land-use decisions using large datasets to estimate discrete choice models. Prior to that, he earned a double major in mathematics, education, and physics from Andrews University in Michigan. His work is concerned with harnessing the geographic information system and parallelized computation to help deal with research problems of a spatial nature or those which are prohibitively difficult to tackle using conventional computational resources. The interactions among land surface, water, and human activity usually fall into these categories. His publications include *Climate Change. Impact on Agriculture and Costs of Adaptation,* co-authored with G. Nelson, M. W. Rosegrant, J. Koo, T. Sulser, T. Zhu, C. Ringler, S. Msangi, A. Palazzo, M. Batka, M. Magalhaes, R. Valmonte-Santos, M. Ewing, and D. Lee (IFPRI); *Building Climate Resilience in the Agriculture Sector* (Asian Development Bank and IFPRI); and *Food Security, Farming, and Climate Change to 2050,* co-authored with G. Nelson, M. W. Rosegrant, A. Palazzo, I. Gray, C. Ingersoll, S. Tokgoz, T. Zhu, T. B. Sulser, C. Ringler, S. Msangi, and L. You (IFPRI).

Myles Fisher (mylesjfisher@gmail.com) is an emeritus scientist at the Centro Internacional de Agricultura Tropical (CIAT) in Cali, Colombia. He worked on the agronomy and physiology of tropical pastures in the Northern Territory of Australia and southeast Queensland with the Commonwealth Scientific and Industrial Research Organisation and in Colombia at CIAT. He was lead scientist for the CGIAR InterCenter Working Group on Climate Change. He consults with CIAT on simulation modeling, soil carbon dynamics, and climate change, and with the United Nations Development Programme on the preservation of Tajikistan's (Central Asia) agrobiodiversity in the face of climate change. Fisher received his bachelor's degree from the University of Melbourne and his PhD from the University of Queensland. He has published more than 100 refereed papers and book chapters and recently edited *Water, Food and Poverty in River Basins: Defining the Limits* (Routledge).

Cindy Cox (c.cox@cgiar.org) is a technical writer at IFPRI. She joined the Environment and Production Technology Division 2011 as a technical writer and researcher for HarvestChoice. She holds a bachelor's degree in biology from Humboldt State University and a master's degree and PhD in plant pathology from Washington State University and Kansas State University, respectively. Before joining IFPRI, she was an agricultural scientist at The Land Institute in Salina, Kansas. Her research interests included the breeding, cytogenetics, and crop protection of perennial grains and the effects of biodiversity on plant diseases and soil microbial communities. Before attending graduate school, Cox was a Peace Corps volunteer in the Central African Republic and taught secondary school mathematics. Her publications include "Spatial Connectedness of Plant Species: Potential Links for Apparent Competition via Plant Diseases" co-authored with W. W. Bockus, R. D. Holt, L. Fang, and K. A. Garrett (in *Plant Pathology* 62: 1195–1204); "Plant Perennials to Save Africa's Soils," co-authored with J. D. Glover and J. P. Reganold (in *Nature* 489: 359–361); and "Progress in Breeding Perennial Grains," co-authored with T. S. Cox, D. L. van Tassel, and L. R. DeHaan (in *Crop and Pasture Science* 61: 1–9).

Karen Garrett (kgarrett@ksu.edu) is a professor in the Department of Plant Pathology at Kansas State University. She holds a bachelor's degree in international agronomy from Purdue University, two master's degrees in plant pathology and statistics from Colorado State University, and a PhD in botany and plant pathology from Oregon State University. Her projects include studies of the effects of environmental trends and variation on plant disease and pests in agricultural and natural systems, strategies for durable disease and pest resistance in crop plants, epidemiological modeling in complex host assemblages, and strategies for sustainable disease and pest management for resource-poor farmers. Garrett has published more than 70 refereed papers and book chapters. The website http://www.ksu.edu/pdecology includes more information about research in her lab. Her publications include "Agricultural Impacts: Big Data Insights into Pest Spread" (in *Nature Climate Change* 3: 955–957); "The Effects of Climate Variability and the Color of Weather Time Series on Agricultural Diseases and Pests, and Decision Making for Their Management," co-authored with A. Dobson, J. Kroschel, B. Natarajan, S. Orlandini, S. Randolph, H. E. Z. Tonnang, and C. Valdivia (in *Agricultural and Forest Meteorology* 170: 216–227); and "Complexity in Climate Change Impacts: A Framework for Analysis of Effects Mediated by Plant Disease," co-authored with G. A. Forbes, S. Savary, P. Skelsey, A. H. Sparks, C. Valdivia, A. H. C. van Bruggen, L. Willocquet, A. Djurle, E. Duveiller, H. Eckersten, S. Pande, C. Vera Cruz, and J. Yuen (in *Plant Pathology* 60: 15–30).

Nicostrato D. Perez (n.perez@cgiar.org) is a senior scientist in the Environment and Production Technology Division, IFPRI, and works on projects related to agricultural growth, sustainable development, and environment-economy modeling. He is part of IFPRI's IMPACT (International Model for Policy Analysis of Agricultural Commodities and Trade) team. Prior to joining IFPRI, he was a professor in the Sustainable Development Program of the Mindanao State University in the Philippines. He holds a PhD in agricultural and applied economics from Virginia Tech, Blacksburg, and a master's degree in agribusiness management from the University of the Philippines at Los Banos. He has worked at three CGIAR centers (IFPRI, the International Rice Research Institute, and WorldFish) and at the Center for Development Research, University of Bonn, Germany. His research interests include environment-economy modeling, water resource management, sustainable development and agricultural policy research.

Pascale Sabbagh (p.sabbagh@cgiar.org) is a senior program manager in the IFPRI-led CGIAR Research Program, Policies Institutions, and Markets. She has worked in research, policy, and project management in both the public and the private sectors in Canada, France, Sweden, and the United States. She has served as a researcher on forestry issues and as chief of the agricultural and forest division in regional administration in France. She has also been a management consultant on organizational and information systems. Sabbagh holds two master's degrees — one in agronomy and one in forest sciences — from Agro Paris Tech in France.

Index

Page numbers for entries occurring in boxes are followed by a *b;* those for entries in figures, by an *f;* those for entries in notes, by an *n;* and those for entries in tables, by a *t.*